AIRPORT DESIGN AND OPERATION

Related Elsevier books

BROOKS	Sea Change in Liner Shipping: Regulation and Managerial Decision-Making in a Global Industry
CAVES & GOSLING	Strategic Airport Planning
HENSHER & BUTTON	Handbook of Logistics and Supply-Chain Management
HENSHER & BUTTON	Handbook of Transport Modelling
HENSHER & BUTTON	Handbook of Transport Systems and Traffic Control
OUM, PARK & ZHANG	Globalization and Strategic Alliances: The Case of the Airline Industry
TURRO	Going Trans-European: Towards a Policy of Building and Financing Transport Networks for Europe

Related Elsevier journals

International Journal of Transport Management: An Official Journal of the Institute of Logistics and Transport

Journal of Air Transport Management
Editor: Ken Button

Transport Policy: Journal of the World Conference on Transport Research Society
Editors: Moshe Ben-Akiva, Yoshitsugu Hayashi and John Preston

Free specimen copies available on request

AIRPORT DESIGN AND OPERATION

ANTONÍN KAZDA
University of Žilina

and

ROBERT E. CAVES
Loughborough University

2000

PERGAMON
An Imprint of Elsevier Science
Amsterdam - Lausanne - New York - Oxford - Shannon - Singapore - Tokyo

ELSEVIER SCIENCE Ltd
The Boulevard, Langford Lane
Kidlington, Oxford OX5 1GB, UK

First edition 2000

Library of Congress Cataloging in Publication Data
A catalog record from the Library of Congress has been applied for.

British Library Cataloguing in Publication Data
A catalogue record from the British Library has been applied for.

ISBN: 0 08 042813 4

♾ The paper used in this publication meets the requirements of ANSI/NISO Z39.48-1992 (Permanence of Paper).
Printed in The Netherlands.

Dedication

We have written this book for all the fools who love the beautiful fragrance of the burnt kerosene.

Tony Kazda and Bob Caves

CONTENTS

PREFACE

This book has the sub-title 'design and operation'. However, the reader will not find chapters devoted exclusively to airport design or airport operation. Airport design and airport operation are closely related and influence each other. A poor design affects the airport operation and results in increasing costs. On the other hand it is difficult to design the airport infrastructure without sound knowledge of the airport operation. This is emphasised throughout the book.

The book does not offer a set of simple instructions for solutions to particular problems. Every airport is unique and a simple generic solution does not exist. Some of the differences that relate to the political and economic situations in Eastern and Western Europe are reflected here. The book explains principles and relationships important for the design of airport facilities, for airport management and for the safe and efficient control of operations. We hope that we have been able to overcome the traditional view that an airport is only the runway and tarmac. An airport is a complex system of facilities and usually the most important enterprise of a region. It is an economic generator and catalyst in its catchment area. However, this book is focused on one narrow part of the airport problem, namely design and operation, while bearing the other aspects in mind. It is up to the reader to judge how successful we have been.

We would like to thank our wives for understanding during our writing, because the time involved for this work was stolen from our families.

Tony Kazda and Bob Caves

Zilina, Slovakia and Loughborough UK, April 2000

1

AIR TRANSPORT AND AIRPORTS

Tony Kazda and Bob Caves

1.1   DEVELOPMENT OF AIRPORTS

First, consider the well-known question: 'What was there earlier?' In the context of this book, it does not refer to the notorious problem about a chicken and an egg, but about an airport and an aeroplane. In fact, the answer is clear. The aeroplane came first. When aviation was in its infancy, the aviator first constructed an aeroplane, and then began to search for a suitable 'airfield', where he could test the machine. The aerodrome parameters had to be selected on the basis of performance and geometrical characteristics of the aircraft. That trend to accommodate the needs of the aircraft prevailed, with some notable exceptions like New York's la Guardia airport, until the end of the 1970s. This was despite the increasing requirements for strength of pavements, width and length of runways and other physical characteristics and equipment of aerodromes. The aerodromes always had to adapt to the aircraft.

The first aeroplanes were light, with a tail wheel, and the engine power was usually low. A mowed meadow with good water drainage was sufficient as an aerodrome for those aeroplanes. The difficulty in controlling the flight path of these aeroplanes required the surrounding airspace to be free of obstacles over a relatively wide area. Since the first aeroplanes were very sensitive to cross wind, the principal requirement was to allow taking off and landing always to be into wind. In the majority of cases, the aerodrome used to be square or circular without the runway being marked out. The wind direction indicator that was so necessary in those days still has to be installed at every aerodrome today, though its use now at big international airports is less obvious. Other visual aids that date from that period are the landing direction indicator and the boundary markers. The latter aid determined unambiguously where the field was, and where the aerodrome was, this flight information for the pilot not always being evident in the terrain.

Immediately after World War I in 1919-1920, the first air carriers opened regular air services between Paris and London, Amsterdam and London, Prague and Paris, among others. However, in that period no noticeable changes occurred in the airport equipment, or in the basic operating concept.

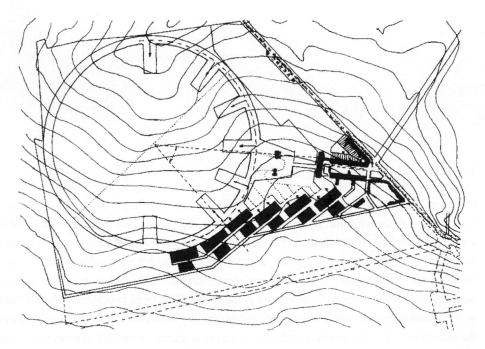

Figure 1-1   The design of Praha-Ruzyně airport development prepared for the competition in the year 1931, gained 2ⁿᵈ price (source: Czech Airports Authority)

Even in the 1930s, the new technology of the Douglas DC-2 and DC-3, which were first put into airline service in 1934 and 1936 respectively, was not sufficiently different to require large changes in the physical characteristics of aerodromes, so the development of airports up to that period may be characterised as gradual. The first passengers on scheduled airlines were mostly business people or the rich and famous, but this was a small scale activity, most of the flying being done by the military. The main change in the airfield's physical characteristics was the runway length. The multiengine aircraft required the length to increase to approximately 1 000 m.

The increasing number of aircraft, and the training of the military pilots required more support facilities at airfields, such as hangars, workshops and barracks.

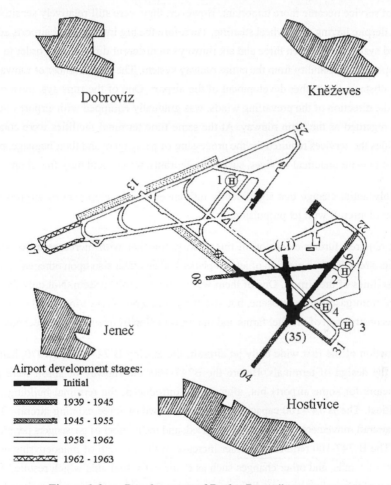

Figure 1-2   Development of Praha-Ruzyně runway system

War does not benefit mankind but, for aviation, it has always meant a rapid step change in development. After World War II, there were unusually favourable conditions for the development of civil aviation and air transport. On one hand there were damaged ground communications, while on the other hand, there were plenty of surplus former military aircraft. There was also the requirement to support the supply chains from the USA to Latin America, to Japan and to Europe under the Marshall Plan. All of that activity allowed civil air transport to recover quickly and then to continue to a higher level than before World War II. The requirements for aerodromes changed dramatically in that same short period of time.

The new aircraft required paved runways, partly because they were heavier and partly because regularity of service became more important. However, they were still relatively sensitive to the crosswind, despite having nose-wheel steering. Therefore the big international airports adopted a complicated system of between three and six runways in different directions in order to provide sufficient operational usability from the entire runway system. The large number of runways often became an obstacle for further development of the airport. One of the runways, most often the runway in the direction of the prevailing winds, was gradually equipped with airport visual aids, thus being regarded as the main runway. At the same time terminal facilities were constructed which, besides the services required for the processing of passengers and their baggage, provided also the first non–aeronautical services, such as restaurants, toilets, and duty free shops.

The next substantial change that significantly influenced the development of airports was the introduction of aircraft with jet propulsion.

Jet aircraft required further extension of the runway, together with increases in its width and upgrading its strength. The operation of jet aeroplanes had an effect also upon other equipment and technical facilities of the airport. One of them was the fuel supply system. Not only did the fuel type change from gasoline to kerosene, but also the volume per aircraft increased considerably, requiring reconstruction of the fuel farms and the introduction of new refueling technologies.

The introduction of the first wide body jet aircraft, the Boeing B 747-100 in 1970, had a large impact on the design of terminals. Before the B747-100, the runway or apron were limiting capacity factors for some airports but, after it was introduced, the terminal building capacity became critical. The B 747-100 capacity could replace two or three existing aircraft. Thus the number of aircraft movements was relatively reduced, and the number of passengers per movement increased. The B 747-100 required a further increase in the strength of manoeuvring areas, the enlargement of stands, and other changes such as to airport visual aids which resulted from the greater height of the cockpit giving a different view from the cockpit during approach and landing.

The B 747-100 in fact symbolized a whole new era of widebody air transport, as well as causing the system to adapt to it. At the same time, it signified that there had to be a limit to which airports could adapt fully to whatever the cutting edge of aircraft technology demanded of them. Not only was there a reaction from the international airport community. The manufacturers themselves came to realize that if they constructed an aeroplane with parameters requiring substantial changes of ground equipment, they would find it difficult to sell it in the marketplace. Futuristic studies of new aircraft in the early 1980s, with a capacity of 700-1000 seats were not taken beyond the paper stage, partly for this reason as well as because the airlines had found it hard to sell all the capacity offered by the 747. Following this argument, the Boeing B 777-200 has been designed with folding wingtips, though this option has not yet been taken up by any airline. The Airbus A 3XX study does not contemplate a similar adaptation for the time being, but it has been designed to fit into an 80 m box which the airport industry regarded as the maximum it could cope with economically.

Most recent changes to airports have not been provoked by new aircraft technology, but by political and economic developments. The airport situation in Europe has changed considerably since the 1960s. The airport in the past was a 'shop-window' of the state, and together with the national flag carrier, also an instrument to enforce state policy. After the successful corporatisation and then the privatisation of the British Airport Authority and some other airports, many governments have gradually changed their policy towards airports, particularly in regard to subsidy.

The following important factors influenced the entire development of airports from 1975 to 1992:

1. The threat of terrorism and a fear of unlawful acts.

2. The privatisation of airports.

3. The progressive deregulation of air transport.

The threat of terrorism, and in particular the bomb attack against the B 747 Pan-Am Flight 103 on 23rd December 1988 near Lockerbie in Scotland, subsequently required expensive changes of airport terminal buildings with a consistent separation of the arriving and departing passengers and installation of technical equipment for detecting explosives.

The privatisation of airports started in Great Britain in 1986, and represented a fundamental change in the manner of administering and financing the airports in Europe. It was and still is seen almost unanimously as a success. It has resulted in a considerable extension and improvement of the services provided, particularly for the passengers and other visitors of the airport.

The deregulation that began in the USA in 1978 produced a revolution in the development of that industry. Up to then, air transport had been developing in an ordered fashion. Deregulation represented a free, unlimited access to the market, without any capacity and price limitations, unblocking the previously stringent regulation of the market in the United States. The percentage of the population who had never before traveled by plane reduced from 70 % to 20 %. However, it also brought about negative consequences for airport capacity due to the concentration of traffic at the major hubs and due to the gradual creation of extremely large airlines with the features of strong monopolies.

Therefore in Europe deregulation was approached with considerable caution, to the extent that the term 'liberalisation' has been adopted for the policy. The first measures to affect the major airlines were adopted by the states of the European Twelve in 1988. They referred in particular to the determination of tariffs and the shares of route capacity. They allowed more flexibility and easier access to the market when certain requirements were fulfilled, free access for aircraft of up to 70 seats and conferment of the Fifth Air Freedom within the states of the European Community.

The rate of growth of air transport worldwide since 1990 has been strong. The volume of passengers in regular air transport has doubled in the period from 1990 to 2000, and in the region of the Pacific Basin it has even quadrupled. The air space in Europe is seriously congested. The lack of airspace slots, into which a flight can be accepted by prior arrangement, has been worsening. The queues of aeroplanes have been lengthening, both on the ground and in the air. The costs incurred by delayed flights reach annually USD hundreds of millions.

Table 1-1   World airports ranking by total passengers – 1998 data

Rank	Airport	Total Passengers	% Change
1	Atlanta (ATL)	73 474 298	7.7
2	Chicago (ORD)	72 485 228	3.0
3	Los Angeles (LAX)	61 215 712	1.8
4	London (LHR)	60 659 593	4.3
5	Dallas/Ft Worth Airport (DFW)	60 482 700	n.a.
6	Tokyo (HND)	51 240 704	3.9
7	Frankfurt/Main (FRA)	42 716 270	6.1
8	San Francisco (SFO)	40 060 326	-1.1
9	Paris (CDG)	38 628 926	9.5
10	Denver (DEN)	36 831 400	5.3

Total Passengers: Arriving + departing passengers + direct transit passengers counted once.
n.a. - data not available
Source: ACI Traffic Data

Besides the need of funding for reconstruction and the building of new terminals, the biggest problem for many large European airports is the lack of capacity of the runway system, leading to a requirement for new runway construction. This is accentuated by the development of regional transport which will continue throughout Europe. Regional transport serves mostly business travellers for one-day trips or to feed long haul flights, thereby increasing the demand for capacity of runway systems during the peak hour. It is possible, as in the USA, to make use of the different characteristics of regional transport aircraft to implement a separate system of approach and take-off but, according to the International Civil Aviation Organisation (ICAO), 16 European airports have insufficient capacity in the year 2000. It is impossible to adopt a quick and effective solution in Europe, the construction of new capacity being hindered by the legal procedure to which projects should be submitted for public discussion in most countries.

The changing structure of air transport, including not only the increasing number of small aircraft intended for direct inter-regional transport, but also the trend to liberalisation and the universally growing transport volumes, will even further increase the pressure on airport capacity. In addition, the airports must also satisfy the changing profile and new categories of passengers. They must prepare for increasing numbers of elderly people, of young parents with children and of the disabled. New standards have made it necessary to reconstruct completely some airport terminals. All these pressures will cause substantial increases in investment. Similar changes will appear in the carriage of freight.

The process of major airport development is taking progressively longer. The second Munich airport was only opened 30 years after the first plans were drawn up. It appears that any approval for the Fifth terminal at Heathrow could only open 12 years after the inquiry into it began and nearly 25 years after it was recognised as being necessary. It is therefore sensible to predict requirements perhaps 35 years ahead, yet the ability to predict even 15 years ahead is questionable.

The Far East is likely to be the most rapidly expanding region from the viewpoint of further development of airports as the economy recovers its former vitality. During the next 20 years it may be foreseen that at least six big international airports will be built there. The construction of each of them will take 10 years as a minimum, but with some of them, it will take up to 20 years. The airports at Osaka Kansai, Hong-Kong Chek Lap Kok, Macau and Kuala Lumpur were opened in the 1990s.

After the opening of Munich II, Oslo Gardermoen and Milan Malpensa, the opening of Athens Sparta in the near future, and the probability of a replacement airport for Lisbon, the development of the network of new international airports in Western Europe may be considered as almost complete. The remaining option for increasing capacity is the development of existing airports. In general, this will be problematic because they have not reserved sufficient land for further development. The biggest problems may be expected in the London area. Although the construction of Terminal 5 is planned for Heathrow Airport, it will not solve the problem of the runway system limiting capacity. The BAA will try to get permission for another runway at one of its London airports, as will Frankfurt, but it will not be easy. In contrast, Charles de Gaulle (Roissy) has considerable room for development and is in the process of opening two new runways.

The airports in the countries of the former Soviet block do have plenty of room for expansion. The relatively uncongested air space and the very bad condition of the ground transport infrastructure point to air transport as the only way to access these countries in the next 15 to 20 years. In comparison with other means of transport, investments in the infrastructure of air transport are small and can be made relatively quickly. The main problem is funding, because the whole Eastern block suffers lack of capital resources.

It is anticipated that the 'hub and spoke' system will continue to be supported in the United States, so causing further pressure on capacity. The latest National Plan for Integrated Airport Systems anticipated a spend of $US 30 billion over five years, including the final elements of the first phase of the new Denver airport. This is the biggest project in the history of airports, being planned to have 16 runways in its final form. In the next decade at least two further major airports are planned in the US, but it is proving difficult to achieve local agreement on the best solution.

These existing and new airports will have to cope with the traffic in 30 and 40 years time. Yet it is not possible to foresee the changes that may occur in management and technology by then. There may be new types of supersonic aircraft. The growing Pacific Basin seems to be the most suitable area from that point of view. With a capacity greater than 200 passengers and flight range over 12 000 km, a supersonic transport would allow a considerable saving of time for passengers and contribute to a dynamic economic growth in the Pacific area. Its introduction would have a major impact on the airport infrastructure. So too will the new large aircraft, especially the Airbus Industrie A 3XX, and even more so its possible successor in the form of a 1 000 seat blended wing-body aircraft.

Development of Heathrow Airport

The development of Heathrow airport is used as an example to illustrate how requirements for the runway system and the other infrastructure of the airport can change rapidly and unexpectedly.

The history of Heathrow airport began in 1929. Richard Fairey Great West Aerodrome, which was used mostly for experimental flights was opened on the site of the present airport. In the course of

World War II the Ministry of Aviation needed to build a bigger airport in the London area with longer runways that could be used by heavy bombers and airliners. In 1942 site selection started and in 1944 it was decided to build an airport at Heathrow with the then classical arrangement of three runways forming an equilateral triangle.

The war terminated before the airport was completed. It was necessary to adapt the airport project to the needs of civil aviation. It was not a simple task. It was necessary to estimate the development of civil aviation and its requirements after a six-year stagnation. A commission of experts assessed several options for completing the construction of the runway system using the three runways under construction. Apart from others, the commission determined these requirements:

1. The runway system should allow the operation of any type of aircraft, considering a crosswind limit of 4 kts (2m.s^{-1}).

2. Two parallel runways should be constructed in each direction, with a minimum separation of 1 500 yards (1 371 m).

The resulting design was in the form of a Star of David. Originally the construction of a third runway triangle was planned to the north of the present airport, beyond the A4 trunk road. Thus a system of three parallel runways would have been available whatever the wind direction. That concept was rejected in 1952. By the end of 1945 the construction of the first runway and several buildings were complete, and Heathrow airport was officially opened on 31st May 1946.

The dynamic developments in air transport made possible by new types of aircraft required constant changes to the original project. By the end of 1947 the construction of the first runway 'triangle' was completed. Work on the second 'triangle' continued simultaneously with the construction of an access tunnel into the central area under runway No. 1. In 1950 the construction of the runway system was practically completed. In order that space may be found for a construction of terminals, apron and the remaining infrastructure in the central area, runway No. 3 and subsequently also other runways were closed very quickly, and thus only three runways are in operation at present (see Figure 1-3).

The original terminal buildings were only of a temporary nature. They were located to the north of the northern runway, and it was clear that in the future they would have to be substituted by a new complex in the middle of the runway system. The construction of the new complex began in 1950 with the control tower and terminal building designed for short routes, this becoming the present Terminal 2. It was completed and opened in 1955.

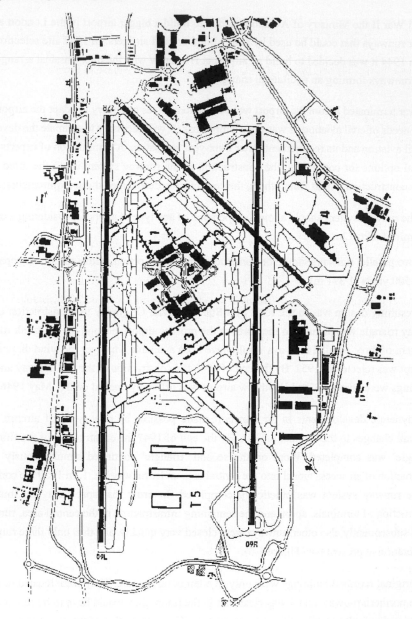

Figure 1-3   London Heathrow airport, original and present runway system and airport terminals

In step with the increasing demand for air transport, new terminal buildings were built. In 1962 Terminal 3 was opened, designed specifically for long haul flights. In 1968 Terminal 1 designed for domestic airlines was opened, and in 1986 Terminal 4, into which all British Airways' longhaul routes were moved from Terminal 3.

This broke out of the central area for the first time, despite the difficulties caused by aircraft having to cross the southern runway and by passengers and bags having to be transferred between Terminal 4 and the central area.

Further increase in traffic beyond the present 60 million passengers per year is limited by the capacity of terminal buildings and stands. Therefore the British Airports Authority has brought forward plans for a fifth Terminal 5 in place of the Perry Oaks sewage farm, between two main runways at the west end of the site. After the public inquiry, which began in 1994, it was expected that the government's consent with the construction would be given in 1997. The first stage of the construction was supposed to last four years, giving a capacity of 10 millions of passengers per year (mppa) and a final capacity of 30 mppa. In fact, as of writing in the year 2000, it is likely that there will be no decision for at least two more years. The likely result will be that some additional capacity will be allowed in exchange for a range of environmental safeguards, and it will probably be necessary to construct the full terminal as soon as possible in order to take the pressure off the existing facilities.

The case for Terminal 5 depends on increases in aircraft size rather than on more flights, but further growth of Heathrow traffic after that would probably require the construction of a new runway.

1.2   STANDARDS

1.2.1   ICAO Legislation

Safety is the overriding requirement in aviation. Standardisation is one of the means to achieve it. In the case of airports, it is standardisation of facilities, ground equipment and procedures. The only justification for differences is to match the types of aircraft that may be expected to use the airports. It is, of course, necessary for the standards to be appropriate and to be agreed by the aviation community.

Although attempts to reach agreement had been made much earlier, the need to agree common requirements for airports used by air carriers became more pressing after World War II. In compliance with Article 37 of the Convention on International Civil Aviation in Chicago in1944, the International Civil Aviation Organisation (ICAO) adopted Annex 14-Aerodromes to the Convention on 29th May 1951. Annex 14 provides the required set of standards for aerodromes used by international air transport. The Annex contains information for planning, designing and operating airports. With the developments in aircraft technology described in the previous section, together with the consequent changes to airports, Annex 14 has been regularly amended and supplemented. Particular Amendments were in the majority of cases approved at sessions of the respective specialist ICAO conference on Aerodromes, Air Routes and Ground Aids (AGA). Each of the ICAO member states may propose a supplement or amendment to an Annex through its aviation authority. The proposal is usually assessed or further examined by a panel of experts. Each of the member states may nominate its experts to the panel. Within ICAO there are panels that have been dedicated to several specific issues for a long time, e.g.:

AWOP   All Weather Operations Panel–issues of operations under restricted meteorological conditions

VAP   Visual Aids Panel–visual aids of airports

OCP   Obstacle Clearance Panel

Other panels have been formed to consider a specific one-off problem, e.g.

ARCP   Aerodrome Reference Code Panel–method for interrelating specifications of airports

HOP   Helicopter Operations Panel–operation of helicopters.

The conclusions reached by the panels are reported in the form of working papers that are sent to the states for comments. Then amendments and supplements to the Annex are usually approved at the Air Navigation Conference or at the AGA conferences.

Each of the ICAO member states is obliged to issue a national set of Standards and Recommended Practices regulating the points in question for their international airports, and amplifying them as necessary. This can give rise to problems of language. The options for an ICAO member state are either to adopt one of the official ICAO languages (English, French, Spanish or Russian), or to translate it into its own language and notify ICAO accordingly. If there is a need, the member state may adapt some of the provisions in its national Standards ands Recommended Practices if it files the differences with ICAO. The provisions in the Annex have two different levels of obligation and relevance:

Standards contain specifications for some physical characteristics, configuration, materials, performance, personnel or procedures. Their uniform acceptance is unconditional in order to ensure safety or regularity of international air navigation. In the event that a member state cannot accept the standard, it is compulsory to notify the ICAO Council of a difference between the national standard and the binding provision.

Recommendations include specifications referring to other physical characteristics, configuration, materials, performance, personnel or procedures. Their acceptance is considered as desirable in the interest of safety, regularity or economy of international air navigation. The member states should endeavour, in compliance with the Convention, to incorporate them into national regulations. The member states are not obliged to notify the differences between recommendations in the Annex and the national Standards and recommend practices. However it is considered helpful to do so, provided such a provision is important to the safety of air transport.

Furthermore the member states are invited to inform ICAO of any other changes that may occur. In addition, the states should publish the differences between their national regulation and the Annex by the means of the Flight Information Service.

Notes are only of an informative character and supplement or explain in more detail the Standards and Recommendations.

At present Annex 14 has two volumes. Volume I Aerodrome Design and Operations and Volume II Heliports. 4 Besides the Annexes ICAO issues other publications. The following manuals, which supplement Annex 14, include guidelines for aerodrome design, construction, planning and operations.

Aerodrome Design Manual (Doc 9157)

Part 1 - Runways

Part 2 - Taxiways, Aprons and Holding Bays

Part 3 - Pavements

Part 4 - Visual Aids

Part 5 - Electrical Systems

Airport Planning Manual (Doc 9184)

Part 1 - Master Planning

Part 2 - Land Use and Environmental Control

Part 3 - Guidelines for Consultant/Construction Services

Airport Services Manual (Doc 9137)

Part 1 - Rescue and Fire Fighting

Part 2 - Pavement Surface Conditions

Part 3 - Bird Control and Reduction

Part 4 - Fog Dispersal (withdrawn)

Part 5 - Removal of Disabled Aircraft

Part 6 - Control of Obstacles

Part 7 - Airport Emergency Planning

Part 8 - Airport Operational Services

Part 9 - Airport Maintenance Practices

Heliport Manual (Doc 9261)

Stolport Manual (Doc 9150)

Manual on the ICAO Bird Strike Information System (IBIS) (Doc 9332)

Manual of Surface Movement Guidance and Control Systems (SMGCS) (Doc 9476)

The book uses these ICAO documents, which are available from the world regional distribution centres, as primary references. It is not considered necessary to repeatedly refer to them in the text.

1.2.2   National Standards and Recommended Practices

Some countries, like the United States, generate a full set of their own standards and recommendations which complement and expand on those contained in the ICAO documentation. These are published as Federal Aviation Administration (FAA) Advisory Circulars 139 and 150. Many other countries find these useful as reference material.

All signatory countries to ICAO are obliged to apply the ICAO standards to their international airports. It would be uneconomic to apply them fully to their more numerous domestic airports, though it is sensible to take note of the principles embodied in the ICAO documents. Therefore every state has a possibility of making its own national Standards and Recommended Practices dealing with specific problems of domestic airports and airfields exclusively within the territory of the particular state for aerial works in agriculture, general aviation airports as well as for limited commercial operations. Besides the various types of civilian airports, there are also military airports. Their physical characteristics, marking and equipment may be different from the characteristics recommended for civil aerodromes. In creating a national set of Standards and Recommended Practices that does not derive directly from Annex 14 or another ICAO publication, the aviation authority usually puts an expert in charge of elaborating a draft of the document. The document draft is distributed for comments from selected organizations and panels of experts. After inclusion of the comments, the new draft is once more discussed in a wider forum. The proposal is also assessed in relation to other Standards and Recommended Practices. Elaboration of each legal document requires a considerable amount of time and effort. If the necessary amount of attention, together with adequate legal and technical resources are not put into the elaboration of the standard, the consequences can be serious. The document should be supplemented, amended, re-elaborated and exceptions from it should be noted, all of which takes further time and effort.

1.3   AIRPORT DEVELOPMENT PLANNING

The rapid development of air transport in the 1980s caused the capacities of many big European airports to be fully taken up in a very short time. The increasing volumes of passengers and freight will continue making demands for the expansion of airport facilities.

Although the majority of European airports have still an excess of capacity, as concluded at the meeting of the ECAC ministers of transport in 1992, each state is obliged to ensure development of ground infrastructure, to detect and eliminate bottlenecks that are limiting the capacity of the airport system. In that way it should be possible to ensure that the increased requirements for capacity of airports in the future will be met. The solutions to these capacity issues can only be approached successfully on a system basis. It is necessary to assess the capacity of each part of the airport system individually: runway, taxiway system and configuration of apron, service roads, parking lots, cargo terminal and ground access to the airport. The result of such a system study is a proposal for staging the development of airport facilities, elaborated in a Master Plan of the airport.

An airport Master Plan represents a guide as to how the airport development should be provided to meet the foreseen demand while maximising and preserving the ultimate capacity of the site. In the majority of cases it is not possible to recommend one specific dogmatic solution. It is always necessary to search for alternative solutions. The result is a compromise which, however, must never be allowed to lower safety standards.

Planning of an airport's development is usually complicated by considerable differences between types of equipment and the level of the technology of the installations that are required for ramp, passenger and freight handling, and operations on the taxiways and runways.

The Master Plan of an airport may be characterised as: '**a plan for the airport construction that considers the possibilities of maximum development of the airport in the given locality. The Master Plan of an airport may be elaborated for an existing airport as well as for an entirely new one, regardless of the size of the airport**'. It is necessary to include not only the space of the airport itself and its facilities, but also other land and communities in its vicinity that are affected by the airport equipment and activities.

It must be highlighted that a Master Plan is only a guide for:

1/ development of facilities

2/ development and use of land in the airport vicinity

3/ determination of impacts of the airport development on the environment

4/ determination of requirements for ground access.

It is necessary to actually construct each of the planned facilities only when an increasing volume of traffic justifies it. Therefore the Master Plan of an airport should include the plan of the phasing of the stages of building. Table 1-2 shows what may be included in the Master Plan of an airport.

As it has been already emphasized, the Master Plan of an airport is only a guideline, and not a program of construction. Therefore it does not solve details of design. In a financial plan, which is included in Master Plan, it is only possible to make approximate analyses of alternatives for development, though costs of construction over the short term do need to be estimated with some accuracy if decisions are to be made on the economic feasibility. The Master Plan determines the strategy of development but not a detailed plan of how to ensure financing of each of the construction stages.

The basic objective of the elaboration of a Master Plan should be that the interested parties must approve all approve it and the public should accept it.

Table 1–2   Purpose of an Airport Master Plan

I.   General part

    A. An airport master plan is a guide for:
- development of airport facilities, for aeronautical and non-aeronautical services
- development of land uses for adjacent areas
- environmental impact assessment
- establishing of access requirements for the airport

    B. Beside others an airport master plan is used to:
- provide guidance for long and short term planning
- identify potential problems and opportunities
- be a tool for financial planning
- serve as basis for negotiation between the airport management and concessionaires

✈ for communication with local authorities and communities

II.   Types of actions during the airport master planning

A. Policy/co-ordinative planning:
   ✈ setting project objectives and aims
   ✈ preparing project work programs, schedules and budgets
   ✈ preparing an evaluation and decision format
   ✈ establishing co-ordination and monitoring procedures
   ✈ establishing data management and information system

B. Economic planning
   ✈ preparing market outlooks and market forecasts
   ✈ determining cost benefit alternatives
   ✈ preparing assessment of catchment area impact study in alternatives

C. Physical planning
   ✈ system of air traffic control and airspace organization
   ✈ airfield configuration including approach zones
   ✈ terminal complex
   ✈ utility communication network and circulation
   ✈ supporting and service facilities
   ✈ ground access system
   ✈ over-all land use patterns

D. Environmental planning
   ✈ preparing of an environmental impact airport assessment
   ✈ project development of the impact area
   ✈ determining neighbouring communities attitudes and opinions

E. Financial planning
   ✈ determining of airport development financing
   ✈ preparing financial feasibility study in alternatives
   ✈ preparing preliminary financial plans for the finally approved project alternative

III.   Steps in the planning process

A. Preparing a master planning work programme

B. Inventory and documentation of existing conditions

C. Future air traffic demand forecast

D. Determining gross facility requirements and preliminary time-phased development of same

E. Assessing existing and potential constraints

F. Agree upon relative importance or priority of various elements:
    - ✈ airport type
    - ✈ constraints
    - ✈ political and other considerations

G. Development of several conceptual or master plan alternatives for purpose of comparative analysis

H. Review and screen alternative conceptual plans. Provide all interested parties with an opportunity to test each alternative

I. Selection of the preferred alternative, development of this alternative and preparing it in final form

IV.   Plan update recommendations

A. Master plan and/or specific elements should be reviewed at least biennially and adjusted as appropriate to reflect conditions at the time of review

B. Master plan should be thoroughly evaluated and modified every five years, or more often if changes in economic, operational, environmental and financial conditions indicate an earlier need for such revision

Source: Airport Planning Manual, Part 1 Master Planning, (ICAO Doc 9184-AN/902)

2

RUNWAYS

Tony Kazda and Bob Caves

2.1   AERODROME REFERENCE CODE

As with other Annexes, the main objective of Annex 14 is to determine criteria for providing world-wide security, regularity and economy of international air transport.

Many provisions which are included in Annex 14 are of universal character and are applicable to the majority of personnel in international air transport, its procedures and equipment. However, some of the provisions are applicable only to specific types of operation, or depend on critical dimensions or performance characteristics of the given aeroplane.

Therefore the main task is to achieve an effective and efficient relationship between particular facilities and aeroplanes. Since there are many types and models of aeroplanes in operation, it would be practically impossible to develop specific facilities for each of the types of aircraft configurations in common use, with all the variations in size and in numbers and location of engines. In Annex 14, there are therefore many facility sizing criteria related to the most demanding (critical) type of aeroplane or group of aeroplanes that may be used in the airport. The established system of aerodrome reference code (Table 1-2) results from that principle in Annex 14.

Table 2-1   Aerodrome reference code

Code element 1		Code element 2		
Code number	Aeroplane reference field length	Code letter	Wing span	Outer main gear wheel span /ᵃ
1	Less than 800 m	A	Up to but not including 15 m	Up to but not including 4.5 m
2	800 m up to but not including 1 200 m	B	15 m up to but not including 24 m	4.5 m up to but not including 6 m
3	1 200 m up to but not including 1 800 m	C	24 m up to but not including 36 m	6 m up to but not including 9 m
4	1 800 m and over	D	36 m up to but not including 52 m	9 m up to but not including 14 m
		E	52 m up to but not including 65 m	9 m up to but not including 14 m
		F	65 m up to but not including 80 m	14 m up to but not including 16 m

a/ Distance between the outside edges of the main gear wheels.

Source: ICAO Annex 14, Aerodromes, Volume I, Aerodrome Design and Operation

Establishment of the Aerodrome Reference Code in Annex 14 pursues three purposes with regards to:

1. Design   **to provide aerodrome designers with guidelines** on how to plan and design an aerodrome by **relating** rational design criteria with current and future aeroplanes' requirements

2. Standard   **the aerodrome reference code** is a simple method for **interrelating the numerous specifications** concerning the characteristics of aerodromes

3. Operation   the aerodrome reference code serves as an indication of the characteristics of the ground facilities, so the prospective **aircraft operator may contemplate whether aerodrome facilities are suitable** for that aircraft to operate there, provided that Annex 14 was complied with during its construction (this is only valid for movement areas).

When the aerodrome reference code was designed, it was decided that it had to consider the aeroplane operational characteristics and geometrical dimensions.

The following parameters may be given as examples of the operational characteristics:

✈ take-off and landing distances

✈ approach speed

✈ nose wheel lifting speed

✈ glide slope angle

✈ climb rate

✈ taxiing speed

✈ turning radius

✈ engine exhaust jet blast velocity.

Examples of geometrical dimensions of aeroplane:

✈ wing span

✈ distance between the engines and the main undercarriage

✈ outer main gear wheel span

✈ distance between the nose wheel and the main undercarriage

✈ height of pilot's eyes above the ground

✈ aeroplane total length

✈ aeroplane height.

The aerodrome reference code is composed of two elements, and ICAO chose to use take off distance, wing span and gear span as the aeroplane indicators, as follows:

The code element 1 (indicated by numbers 1 to 4) is based on the aeroplane reference field length.[1] It refers to the critical aeroplane's normalised takeoff requirements, and relates in particular to the runway geometry and obstacle clearance limit aspects of the design. However, it does not control the actual required length of the runway. The FAA airport coding system uses the approach speed

---

1 Footnote: Aeroplane reference field length is the minimum length required for the take-off at maximum certificated take-off mass at sea level, standard atmospheric conditions, still air and zero runway slope, as shown in the appropriate aeroplane flight manual prescribed by the certificating authority or equivalent data from the aeroplane manufacturer. Field length means balanced field length (the term balanced field length is explained hereinafter) if applicable, or a take-off distance in other cases. The ICAO documentation provides a listing of Reference Field Lengths.

to represent this dynamic part of the aircraft's characteristics.

The code element 2 (indicated by letters A to F) is based on the aeroplane geometrical characteristics. It shall be determined by a wing span and an outer main gear wheel span. It relates in particular with the provisions referring to taxiways and aprons sizes and spacings.

Some provisions specifying the characteristics of movement areas are determined by a combination of both elements of the aerodrome reference code.

2.2   RUNWAY LENGTH

Determination of the runway length to be provided is one of the most important decisions in designing an aerodrome. The runway length determines the types of aircraft that may use the aerodrome, their allowable take-off weight and hence the distance they may fly.

The number of new aerodromes for which runways have to be designed is limited. In the majority of cases the task is to extend existing runways or to provide supplementary runways. The basic requirements for the runway parameters may be specified from market research into the types of aeroplanes, the networks of the operating airlines and prognoses of further market development at the airport in question.

A runway extension often makes it possible to open the aerodrome to larger aeroplanes, and to longer flight sectors, so extending the market of the aerodrome. There is usually considerable flexibility in how an aircraft can reach its maximum operating weight, between the extremes of maximising fuel load with limited payload and maximising payload at the expense of fuel and hence range. Additionally, it may be that the sector to be flown does not require a full load of fuel, or that the passenger load factor is expected to be low, or there is likely to be little cargo. In these cases, the aircraft will not be at maximum take-off weight, and hence may gain no economic advantage from an extension. The first step in specifying the necessary runway length is to create a list of the aeroplanes that may wish to use the aerodrome after the extension and their likely destinations from the aerodrome. It is advisable to divide the aeroplanes into groups which are characterised by:

✈ take-off mass

✈ payload.

Each of the aeroplane groups requiring approximately the same runway length. In the majority of cases, only a small group of aeroplanes requires the longest class of runway, or even only one aeroplane: the 'critical' aeroplane.

The number of movements of a critical aeroplane is sometimes so small that there is no economic justification for extending the runway so that the runway fulfils its requirements. In that case the operation of a critical type is usually still possible from the existing runway, though with a lowered payload, or reduced flying range, or both at some times of the day and year. However, the airlines will not use such an aerodrome if it would frequently require considerable reduction of the payload or limitation of the flying range.

Table 2-2   Categorisation of the Flight Accident Probability

Category (description)	Probability (per flight)
Frequent (it is probable that the event will happen within an aircraft's service life)	more than 10^{-3}
Reasonable probability (it is not probable that the event will happen, but it may happen within an aircraft's service life)	$10^{-3} - 10^{-5}$
Remote (it is not probable that the event will happen to every aircraft but it may happen several times on the same type of aircraft)	$10^{-5} - 10^{-7}$
Extremely remote (it is possible, but not probable, that the event will happen to several aircraft of the same type)	$10^{-7} - 10^{-9}$
Not probable (practically will not happen)	less then 10^{-9}

A runway extension may also be economically unjustifiable because it is too technically demanding to construct it, or because there are practically immovable obstacles in the take-off and approach areas. In that case, a combination of runway and stopway, or runway and starter strip may allow the same take-off length to be provided. It is clear from the above that the determination of the required runway length is exacting, and any case of runway lengthening should be considered specifically for each case, taking all factors into account. It is necessary to consider not only the future development of the air transport market, but also changes in the prices of land, building materials, labour and economic development as a whole. It is often sensible to buy land and make at least a territorial provision for a prospective extension of the runway. In any case, the overriding criterion is to provide the defined degree of safety of the aerodrome operation.

The runway performance characteristics of the critical type of aeroplane may be obtained in several ways:

✈ using the parameters given in the flight manual of the aircraft type

✈ using the parameters given in the aircraft type's airport planning manual (Aircraft Characteristics for Airport Planning) issued by all manufacturers of commercial aircraft

✈ verifying the characteristics by means of a flight test at specified accident probability based on statistical data (see Table 2-2).

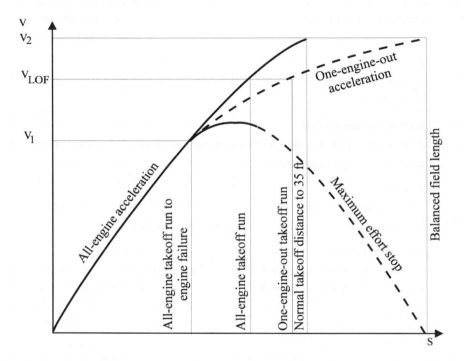

Figure 2-1   Takeoff field length demonstration requirements for multiengine aircraft

From the performance characteristics drawn in the charts given in the flight manual of the critical aircraft type, the landing distance, take-off distance and accelerate stop distance at the various feasible configurations of flap settings and airframe system availability can all be derived. The data are chosen to represent the given parameters of the aerodrome and the proposed operation. The parameters include the aerodrome altitude, air temperature, runway longitudinal slope, as well as the maximum likely aeroplane mass.

The decision as to whether the runway is to be extended or only a stopway and a clearway are to be provided beyond the original runway ends depends, among other things, on the characteristics of the terrain beyond the runway end and on the occurrence of possible obstacles.

The performance characteristics of the critical aeroplane at a particular aerodrome have to allow the aeroplane, with regards to the runway characteristics and actual meteorological conditions, either to complete the take-off safely, or to stop after aborted take-off. For each take-off a speed V_1 is defined, which is called the decision speed (sometimes called also critical speed), see Figure 2-1.

If an engine failure occurs at a speed lower than V_1, the pilot shall abort the take-off. If the aeroplane has a greater speed than V_1 when the emergency is recognised the pilot shall continue the take-off with one engine inoperative. Trying to continue the take-off with an engine failed before V_1 would require a very long take-off distance due to the reduced power available after an engine failure.

On the other hand, there would not have been any problem to stop the aeroplane in the remaining portion of the runway. With an engine becoming inoperative only after exceeding the speed V_1, the aeroplane has a relatively high speed and the remaining power is sufficient for completing the take-off on the remaining runway. After exceeding the speed V_1 before deciding to abort the takeoff the aeroplane speed would be so high that the aeroplane would need an excessive distance to stop safely in the remaining runway.

Decision speed is the speed chosen by the aircraft captain in relation to the respective limitations of the aircraft, the airline operator, runway characteristics and actual meteorological conditions. The accelerate stop distance d_2 (Figure 2-2.) is the sum of the take-off distance to reach the speed V_1 with all engines operating, the distance which the aeroplane travels during the pilot's reaction time and the braking distance until the aircraft comes to a full stop

The take-off length with one engine inoperative, d_3, is the sum of the distance with all engines operating until reaching the speed V_1, the length of a further distance to lift off with one engine inoperative and the climb to the altitude 10.7 m (35 ft) (Figure 2-2).

where the particular speeds are defined as:

V_1 decision speed,

V_{mc} minimum control speed, minimum speed at which engine failure may occur and still a straight flight is possible at this speed and is fully controlled

V_R speed at which the nosewheel lifts

V_{LOF} main gear liftoff (or unstick) speed

V_{mu} minimum unstick speed, >minimum speed which allows safe continuation of takeoff

V_2 safe climbing speed 1.2 V_s, where V_s is a stalling speed for the takeoff configuration.

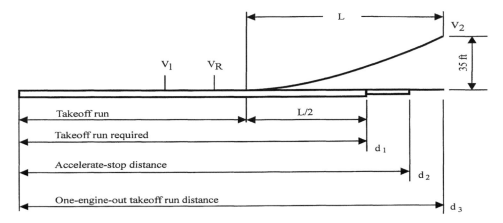

Figure 2-2   One engine-out takeoff/aborted takeoff

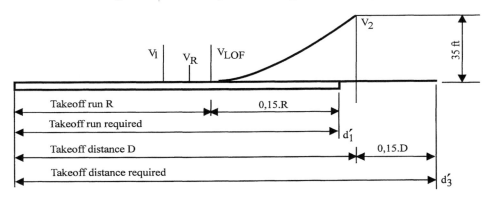

Figure 2-3   Takeoff–all engines

The required safe distance necessary for a take-off is considered to be whichever is the greater of:

→ take-off distance with one engine inoperative - d_3

→ 115 per cent of the demonstrated take-off distance with all engines operating d'_3, (Figure 2.3).

If a lower V_1 is chosen, the take-off distance with one engine inoperative - d_3 (point 1) determines the necessary runway length; while with a higher V_1 chosen, the accelerate stop distance d_2 (point 2) is taken. If the 115 per cent of the take-off distance with all engines operating is so small that it lies under the point 3 (line a), the distance d'_3 has no effect upon the necessary take-off distance.

However, if the 115 per cent of the take-off distance with all engines operating lies over the points 1 and 2 (line c), it is the only determinant for the take-off distance.

If it lies between the points 3 and 1, or 2 (line b), all the above named events may occur. The aircrafts' Airport Planning Manuals normally report only the most dominant of these criteria and only consider the balanced engine-out field length where V_1 is chosen so the distance for take-off and rejected take-off are equal.

For a take-off with one engine inoperative, the hard runway part of the total available distance should be longer than the actual required take-off run. That distance shall be determined by the sum of the aeroplane take-off distance with all engines operating up to the speed V_1, a further distance run with one engine inoperative up to the lift-off speed V_{LOF} and a half of the horizontal distance to climb up to 10.7m (35 ft) (Figure 2-2).

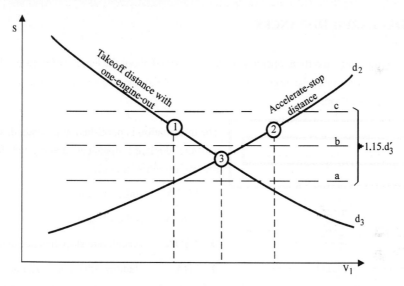

Figure 2-4   *Effect of decision speed V_1 on takeoff distance, accelerate-stop distance and takeoff run*

The distance necessary for a take-off run is considered to be whichever is the greater of:

✈ distance run with one engine inoperative plus half of the horizontal distance to climb to 10.7m (35 ft) - d_1

✈ 115 per cent of the take-off run with all engines operating - d'_1.

In the majority of cases, the aeroplane landing distance is smaller that the take-off distance. The final phase of landing begins at an altitude of 15 m and ends with the full stop of the aeroplane. In practice, the aeroplane often vacates the runway, e.g. through a high-speed exit taxiway, rather than coming to a full stop, but the regulation distances are based on the former case. The necessary runway length for landing should be greater than the actual minimum landing distance demonstrated by the manufacturer, so that day to day variations in pilot behaviour and effects of meteorological conditions may be taken into account. Regulations vary considerably with respect to the margins required over the determined landing distance, depending on the country and the state of the runway, but it is quite common for a margin of 42 per cent to be applied. Most regulations for the determination of landing performance now require the demonstration to be made on a standard wet runway with an appropriate flying technique for the conditions. An additional safety factor is often applied for low visibility landings, which can result in the landing distance being greater than the take-off distance required, particularly for modern twin-engined aircraft.

2.3   DECLARED DISTANCES

In order that the pilots and flight operators may be informed about the lengths of runway, clearway, stopway and runway strip, the airport operator has to publish the respective information in the air information publication.

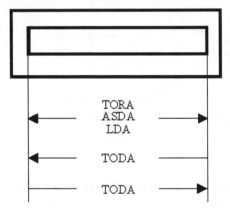

Figure 2-5a   Declared distances

The information is published in a standardised form, and the declared distances determined for each runway direction include:

✈ TORA   (take-off run available)

✈ TODA   (take-off distance available)

✈ ASDA   (accelerate stop distance available)

✈ LDA   (landing distance available).

The particular lengths shall not be determined only by the length of runway or other movement surfaces, but also by their actual condition and contingent obstacles in the take-off or approach areas.

Figure 2-5 a-c gives several examples of declared distances. In general, it may be stated that the take-off run and landing may be performed only on a full strength runway, while remaining part of the take-off may be completed over a runway strip or a clearway.

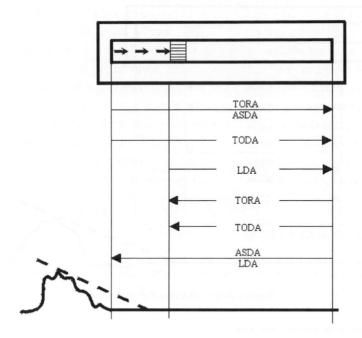

Figure 2-5b   **Declared distances**

In many cases the runway will be full strength for its whole length with thresholds located at the runway ends, runway strips with specified dimensions shall extend beyond the end as well as along each side of the runway. In that case the declared distances will be the same in both directions, see Figure 2-5a.

Alternatively, the runway may have the landing threshold shifted, usually due to obstacles in the approach area. However, the portion of the full strength runway before the displaced threshold may be used for take-off, see Figure 2-5b.

Figure 2-5c illustrates a runway where the threshold was displaced due to a runway section in maintenance (left). Beyond the end of the runway a stopway was established (right). It is not possible to extend the runway in that direction because an obstacle occurs in the take-off area. The stopway is of a full strength and is surrounded by runway strips.

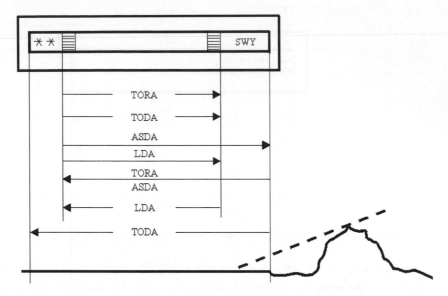

Figure 2-5c   Declared distances

2.4   RUNWAY WIDTH

2.4.1   Runway Width Requirements

As already stated in the above section 2.1 on the Aerodrome Reference Code, the width of a runway is one of the elements that is affected by several geometrical characteristics of aeroplanes:

✈ the distance between the outside edges of the main gear wheels

✈ the distance between wing mounted engines and the longitudinal axis of an aeroplane

✈ the wing span.

However, the required runway width is also affected by the operational elements:

✈ the approach speed of aeroplane

✈ the prevailing meteorological conditions.

Lack of sufficient width will cause constraints on the operations. The minimum runway width is therefore specified in Annex 14 by interrelating both of the code elements, see Table 2–3.

Table 2-3   Minimum runway width

Code number	Code letter					
	A	B	C	D	E	F
1^X	18 m	18 m	23 m	-	-	-
2^X	23 m	23 m	30 m	-	-	-
3	30 m	30 m	30 m	45 m	-	-
4	-	-	45 m	45 m	45 m	60 m

^{X}/ The width of a precision approach runway should be not less than 30 m where the code number is 1 or 2.

Source: ICAO Annex 14, Aerodromes, Volume I, Aerodrome Design and Operation

Under normal conditions, the width of a runway should ensure that an aeroplane does run off from the side of the runway during the take-off or landing, even after a critical engine failure causing the aircraft to yaw towards the failed engine. Deviation from the runway axis during take-off and landing occurs in particular due to a lack of ability in tracking the centreline. With impaired visibility, adverse cross – wind conditions, or poor braking action, the probability of a touch-down, or a slide sideways from the runway axis during take-off increases. Therefore a wider runway is specified for precision approach runways where the code number is 1 and 2.

2.4.2   Runway Edge Strips

According to Annex 14, the width of 45 m is sufficient for runways where the code letter is E, that width of pavement conforming with the undercarriage gauge of the largest aeroplanes. In the event that an aeroplane touches down to the side of the centreline of a runway, even though the undercarriage is on a full strength pavement, the wing mounted engines overlap its edge. Thus they could take in loose material, such as small stones, with a possibility of consequent Foreign Object Damage (FOD) to the engine. Therefore runway shoulders should be provided for a runway where the code letter is D and E so that the over-all width of the runway and its shoulders is not less than 60 m. If the code letter is F, the width of the runway itself must be 60 m.

The shoulders may be turf over stabilised earth, but are usually designed as light asphalt pavements with a load bearing capacity that will support the loads of the ground equipment and reduce the probability of damage to an aircraft veering off the runway. Runway edge lights are located in the runway shoulders. The winter maintenance of the edges of a runway is then considerably simpler than where the lights are installed in the grass strip.

A light asphalt shoulder is not designed for the very rare event of an aeroplane running off the runway, so when it happens it damages the pavement. The repair of an asphalt pavement is quite demanding. Nevertheless, its load bearing capacity should prevent the wheels from being bogged down, which may cause serious damage to the aeroplane. At the same time, a shoulder provides a gradual change of the load bearing capacity between a runway and its associated strip. If a runway is made of cement concrete and a shoulder is made of asphalt, the passage between the runway edge and the shoulder is visually distinctive and clear enough. If both the runway and the shoulder have asphalt surfaces, the edge of the runway should be indicated by a runway side stripe marking. The shoulder should be vertically flush with the runway edge. An example of a runway design with a shoulder is given in the Figure 2-6.

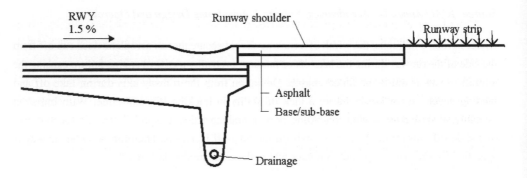

Figure 2-6   Runway with runway shoulder

2.5   RUNWAY SLOPES

2.5.1   Transverse Slopes

Annex 14 recommends that runways be constructed with a roof-like cambered transverse slope, the slope providing quick rainwater drainage particularly at heavy rain and side wind. In spite of that, there are cases when a designer is forced to design a one-sided transverse slope of a runway to reduce costs of earthwork, or to avoid too thick a runway construction when a roof-like profile is selected, or to reduce cost by providing a drainage channel on only one side of the runway. In order that drainage may be ensured, the transverse slope of a runway should be at least 1 per cent but should not exceed 2 per cent, depending on the code letter. The transverse slopes should normally facilitate drainage and minimise the layer of water accumulated on the runway. In some cases the drainage is improved by a transverse grooving of the pavement which, however, in no case may replace the prescribed slopes.

2.5.2   Longitudinal Slopes

Few runways are level throughout their length in the longitudinal direction. Often there are several slope changes in its length. A completely horizontal runway could be constructed, but this would generally require considerable earthwork. On the other hand, there is a limitation on the allowable longitudinal slope due to the performance characteristics of particular types of aeroplanes and the requirements of the operation. The Annex 14 standard therefore limits the so called average slope of a runway, which is calculated by dividing the difference between the highest and the lowest points of a runway by its all-over length; and also limits the maximum slope of any arbitrary portion of a runway, the profiles usually being controlled in segments of 50 to 100 m.

For example, the longitudinal slope of the precision approach runways where the code number is 3 and 4 is limited to 0.8 per cent in the first and last quarter. The strict limitation of the slope is determined either in order to facilitate the final phase of flare in impaired meteorological conditions or by technical parameters of the equipment for a precision instrument approach, particularly the radio–altimeter. The longitudinal slope of non-precision runways of civil airports is limited to 1-2 per cent, depending on the code number.

In special cases, a longitudinal slope up to 8 per cent is permitted for one-way runways in airfields for agricultural activities, for special airfields for mountain rescue services, or, indeed, for commercial service in mountainous areas of the third world.

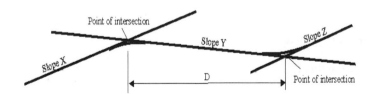

Figure 2-7   Profile of centre line of runway
Source: ICAO Annex 14, Aerodromes, Volume I, Aerodrome Design and Operation

Longitudinal slope changes are also limited. The objective of limiting local roughness is to confine dynamic load of the undercarriage system of an aeroplane when it moves in a high speed on the runway, and to secure permanent and safe contact of the tyres with the surface of the runway. Longer slope change limitations are to ensure the pilot has an adequate view of the runway. A change from one slope to another should be achieved by a minimum radius curvature of 7 500 m - 30 000 m, depending on the code number of the runway. Similarly, the distance between the apexes of two neighbouring curvatures shall be limited. The distance should not be less than the sum of the absolute numerical values of the difference of longitudinal slopes multiplied by a factor, depending on the magnitude of the curvature, in turn depending on the code number of the runway, or 45 m, whichever is the greater (see Figure 2-7).

$$D = k \, [\, |x - y| + |y - z| \,]$$

where:

D   minimum distance between point of intersection of slope changes in metres

k   minimum radius of curvature (between 7 500 – 30 000 with respect to the RWY code number)

x,y,z   longitudinal slopes per mile

The longitudinal slope changes are furthermore limited by the requirement of mutual visibility between two points at a height of 1.5 to 3 m (according to the code number of a runway) and at a distance which is equal to at least one half of the runway length.

3

RUNWAY STRIPS AND OTHER AREAS

Tony Kazda and Bob Caves

3.1 RUNWAY STRIPS

Each runway should be surrounded by a runway strip. The runway strip is intended to ensure the safety of an aeroplane and its occupants in the event of an aeroplane:

✈ running off the runway during landing or takeoff

✈ deviating from the runway centreline during a missed approach.

The physical characteristics and other requirements for runway strips are derived from the need to allow for these events.

The required width of a runway strip depends on the degree of approach guidance to the runway and also on the runway reference code number. The width of a non-instrument runway strip should extend for 30 to 75 m on each side of the centre line of the runway. A strip of an instrument and a precision approach runway should extend for 75 to 150 m. The runway strip should extend past the runway end by 30 to 60 m. No equipment or constructions which may create an obstacle or endanger aeroplanes should be situated on a runway strip. The only exception is radio-navigation, visual and other equipment which should be installed on the runway strip to support the operation of aircraft, for example the Instrument Landing System (ILS) glide path antenna, Precision Approach Radar (PAR), runway visual range (RVR) equipment, and the like. If possible, the equipment should be mounted on frangible fittings.

In the majority of cases the surface of a runway strip should be grass. It is intended that, in the event of an aeroplane running off the runway, the undercarriage of the aeroplane should gradually sink into the runway strip, thus providing effective deceleration of the aeroplane. The aeroplane's structure should not suffer serious damage.

It is probable that in the event of an aeroplane running off the runway, the aeroplane will not need the whole width of the runway strip. Therefore the surface of a runway strip need not have the same quality over all its area. Depending on the category of approach guidance of a runway and its reference code number, the surface of a runway strip should be able to cater for the event of an aeroplane running off to a distance of 30 to 75 m from the centre line of the runway. This part should have a bearing strength of approximately CBR 20, which requires the top 20 cm of soil be compacted. That is not only a question of complying with the specified slopes and quality of the grass surface. Over the whole portion of a runway strip to the specified distance from the runway, there must not be any solid vertical constructions even under the surface of the runway strip to a depth of 30 cm and objects which a sinking wheel of the aeroplane might impact. Such obstacles may, for example, be foundations of lighting installations and airside signs, drainage gutter inlets, or the edges of taxiways within the runway strips. At least the upper 30 cm of such a construction should be chamfered. Other equipment which need not be installed at the surface level should be sunk to a depth of at least 30 cm.

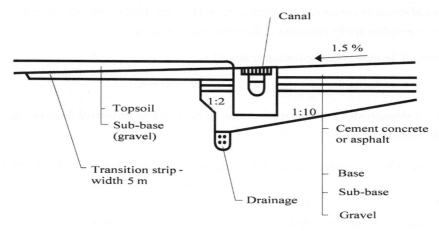

Figure 3-1   Transition strip – gravel and topsoil

The strength of a runway strip will depend on geological conditions and the height of the water table. It may be necessary to put in drains to reclaim the land. In exceptional cases, a runway strip should be strengthened in the same way as a runway with a grass surface at general aviation airports. In addition, it is necessary to guarantee a gradual rate of change in the reduction of strength from the edge of a runway or runway shoulders to facilitate the smooth deceleration. If the edge of a runway is joined directly with a grass strip, it is possible to design wedges of compacted gravel or fine gravel upon which a layer of topsoil is spread for a distance of approximately 5 m from the edge of the runway see Figure 3-1. A gradual change of the strength is provided also by runway shoulders if they are constructed, as described in Chapter 2.

There are firm recommendations for the control of temporary obstacles in the strip. The extent to which hazards may be created by work in progress should be controlled by the relevant authority. It should take account of the type of aircraft, the runway width, the meteorological conditions and the possibility of using other runways. If work is allowed, the resulting hazards must be promulgated by NOTAM. In Zone 1, within 21 m of the runway edge, work may only take place on one side of the runway. The area under work must be of limited size, and the clearance of engines and propellers must not be compromised. No plant or machinery should operate in this area when aircraft are using the runway. If an aircraft is disabled in this zone, the runway must be closed. Less severe restrictions apply in Zone 2, which extends to the edge of the graded portion of the strip.

If an airport is used by aeroplanes with jet engines, the portion of a runway strip within a distance of at least 30 m before a threshold should be sealed over the width of the runway. This prevents erosion by the action of exhaust gases and exposure of an edge of the runway pavement. The construction of a light pavement for such purpose may be similar to that of runway shoulders.

A grass strip for emergency landing of aeroplanes with retracted landing gear should be provided at each airport which has a paved runway. The strip should have dimensions of at least 1 000 m by 100 m, except airports with paved runways where the code number is 1 and 2, where a length of emergency grass strip equal to the runway length is sufficient. The strip for emergency landing can usually be constructed on a runway strip.

As with other airside dimensions, those specified for a runway strip should be understood as the minimum requirements, and in some cases it may be necessary to widen a runway strip. On the other hand, there may be cases where a smaller width of a runway strip is sufficient, for example, for the portion of a runway intended only for take-off, the so called 'starter-strip'.

3.2 CLEARWAYS

A clearway may be provided to extend the takeoff distance available (TODA) beyond the end of the hard surface which defines the declarable length of the available takeoff run (TORA). It implies that the ground in the clearway should not project above a plane having an upward slope of 1.25 per cent between the end of the hard runway and the end of the clearway. A clearway is feasible because the last part of a takeoff takes place in flight rather than on the ground. It is usual to also provide a stopway, though often rather shorter than the clearway, since aircraft cannot normally use a very unbalanced takeoff. To the extent that the aircraft's performance characteristics permit it, the clearway allows a pilot to choose a relatively low decision speed, after which it would be advisable to fly rather than stop if an emergency occurred. Then a relatively shorter accelerate-stop distance is needed and hence a lower cost of prepared surface. A clearway should extend to a width of at least 75 m on each side of the centre line of the runway and its length should not exceed half the length of the runway length available for take-off. It is subject to the same possible limitation as a more normal provision of TODA, that the imaginary takeoff surface which starts at the end of the clearway must not be broken by obstructions.

3.3   RUNWAY END SAFETY AREAS

Operational experience and statistical data of accidents where the aircraft has landed short or overshot a runway have shown that the portion of the runway strip which is located off the ends of runways does not provide sufficient protection for aircraft in these circumstances. Some of the most serious overrun accidents occur in the event of an aeroplane running off the runway end when the decision to abort a takeoff has been taken after the speed has already exceeded the V_1 speed, after which takeoff should normally be continued. Such cases occur mostly when a pilot judges that, even at that speed, the takeoff cannot be continued because of the nature of an engine failure, or serious vibrations, or when a failure occurs after the normal liftoff speed (V_{LOF}) which makes it impossible to fly, such as an incorrectly adjusted stabiliser trim setting. In that case the aeroplane has considerably greater speed than for example in the event of an aeroplane running off beyond the runway end after landing under poor braking conditions, or when a pilot touches down well beyond the touchdown zone, which is normally some 300 m from the threshold. For those reasons, a runway end safety area (RESA) of at least 90 m should be provided at each end of a runway strip of the runway where the reference code number is 3 or 4 and the runway is an instrument one. The latest edition of Annex 14 recommends that, where possible, 240 m of RESA should be provided for code 3 and 4 runways, making a total of 300 m available to contain an overrun.

Dimensions of runway end safety areas should be determined on the basis of statistical data on an occurrence of accidents beyond the runway end. The ICAO accident data base shows that about eighty per cent of all overrun and undershoot accidents would be contained within a runway end safety area extending 90 m beyond the end of the runway and double the width of the runway, surrounded by a normal runway strip. Ninety per cent of all such accidents would be contained within a runway end safety area 150 m wide and 300 m long. However, The data only give the actual distance that aircraft have overshot the runway, regardless of the available length, the expected distance required for the take-off or landing, and the characteristics of the runway and overrun area. Figure 3-2a,b shows the difference between the ICAO data and the real extent of the excess distance used up in a small sample of landing accidents.

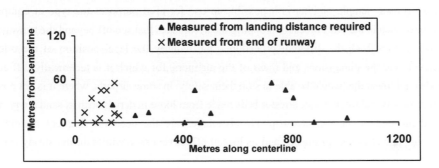

Figure 3-2a   Wreckage location relative to runway end (landing)

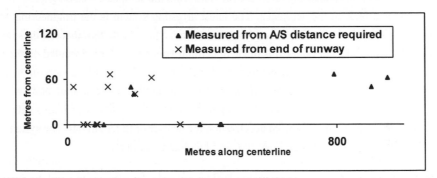

Figure 3-2b   Wreckage location relative to landing distance required

Source: Kirkland, I. and R. E. Caves, (1998), Risk of overruns, ACI Europe Airport Safety Symposium, Riga, September

A similar set of results is obtained for rejected take-off overruns. These differences should be taken into account in the assessment of risk at airports, as it is clear that those airports where a large proportion of landings use the full runway length will be more likely to experience an overrun, other things being equal.

Further enlargement of a runway end safety area is practically impossible due to the fact that at the majority of airports, the localiser antenna of the ILS is located not further than 300 m beyond the end of the runway. In the event of an aeroplane running off into the runway end safety area, it should not sustain such damage as would have a serious consequence for the occupants, for example breach of fuel tanks, breakage of fuselage, or fracture of undercarriages. Therefore, in the runway end safety areas, there must not be any embankments or ditches for roads, railways or water courses. There are also limits to the positive and negative slopes which are allowed.

In some cases there is not enough space to provide a runway end safety area without reducing the declared runway length. Neither is it possible to use for civil purposes the type of equipment commonly used in the Air Forces for stopping the aeroplanes that run off beyond the runway end. The reasons are both the different procedures, particularly the legal position of the pilot-in-command, and the dimensions and mass of the airliners, for which it is technically difficult to provide equipment that would be able to stop them safely. In some airports, where it is not possible to build runway end safety areas, arrestor beds made from loose materials such as sand are provided beyond the end of the runway strip. Disadvantages are the need for frequent maintenance (loosening and cleaning) of the bed and its loss of effectiveness in winter during strong frosts.

In other cases a urea foam has been used to form the bed, the strength of which allows movement of fire vehicles while compacting under the pressure from the aeroplane's landing gear, so burying the wheels and braking the aeroplane. The latest arresting system is the Engineering Material Arresting System (EMAS). EMAS is, like the urea bed, a passive system that requires no action by air traffic controllers or pilots. It is made of cellular concrete material installed on the runway overrun to decelerate an aircraft in emergency. When an aircraft is unable to stop on the runway, it rolls into the arrestor bed. It is decelerated by the load applied to the aircraft as its wheels run through the breaking cellular concrete. The depth of arrestor bed gradually increases with distance from the runway, providing increased deceleration for heavier or faster aircraft. The system is FAA approved and is installed at New York JFK and La Guardia airports.

If safety is to be ensured, it is necessary to consider not only the runway strip and the runway end safety areas. Attention must be given also to the near surroundings of the airport, particularly under the approach path. Here there should not be any elevated cement-concrete constructions, sharp embankments for the containment of sewage plants, canals, or transport links. For example, when a Boeing 737 made an emergency approach to the East - Midlands Airport in Great Britain on 8.1.1989, the aeroplane impacted the embankment before the runway where the motorway M-1 runs though a cutting. The severe impact loads broke the fuselage and the aeroplane wreckage blocked the motorway.

Forty-seven people died in the subsequent fire. If the motorway had been covered, the accident probably would not have claimed victims. There is a whole range of airports with similar dangerous obstacles which fall outside the direct responsibility of the airport and its designers. Yet this is the location of most remaining fatal accidents, largely because the on-airport protection and rescue facilities are so good.

It is also the area where third parties are likely to be killed or injured. Some countries therefore require that a public safety zone (PSZ) be established off the end of busy commercial runways with the purpose of limiting the allowable land uses. The UK has recently adopted a new form of PSZ based on research which shows the area of constant risk to be described by contours of similar shape to the noise contours. They have been approximated to thin triangles in the UK regulations, where new dwellings will not be permitted if the individual risk death is greater than one in 100 000 per year.

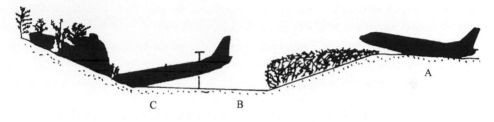

Figure 3-3   ***Phases of East-Midlands accident:***

A/ The first touch before motorway
B/ Impact on the opposite bank
C/ Final position of the destroyed fuselage

4

TAXIWAYS

Tony Kazda and Bob Caves

4.1 TAXIWAY SYSTEM DESIGN

It is often difficult to design an optimum system of taxiways. The taxiway system may have a decisive influence on the capacity of the runway system, and thereby also the overall capacity of the aerodrome. Yet the surface area of taxiways may be greater than the area of runways. For example, in the case of the Osaka's Kansai International Airport, opened in 1994, the taxiways have an overall length of 11.3 km, and the runway have a length of 3 500 m. Considering that the load bearing strength of a taxiway should be equal to or greater than the load bearing strength of a runway, the construction of taxiways represents an important item in the total investment costs. Therefore it is necessary to optimise the taxiway system layout to provide efficient taxiing without undue expense.

The taxiways should permit safe, fluent and expeditious movement of aircraft. They should provide the shortest and most expeditious connection of the runway with the apron and other areas in the airport. This also minimises the fuel consumption of the aeroplanes, which has a positive effect upon the environment. The safety of aeroplanes is enhanced if the taxiways are designed as one-way operation, and crossing other taxiways, and particularly runways, is minimised. In those aerodromes where the number of aircraft movements during the peak hour traffic is relatively small, it is usually sufficient to provide only a short taxiway at right angles to the runway to connect it to the apron.

To cope with larger aeroplanes, it is then usually necessary to provide additional pavement at the ends of the runway to allow the aircraft to turn round. The runway occupancy time is then increased Figure 4-1.

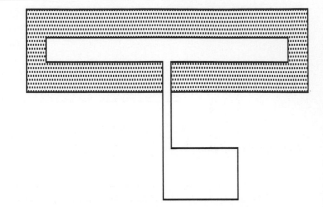

If the number of movements during the peak hour traffic exceeds about 12, consideration may have to be given to construction of a taxiway parallel to the runway, and right angle connecting taxiways to the

Figure 4-1   Runway and apron connected with short right angle taxiway

ends of the runway. In addition, in the event of a longer runway, several right angle connecting taxiways may be constructed, usually in one third or quarter of the runway length.

The system of a parallel taxiway with right angle connections may be sufficient for up to 25 during the peak hour movements, Figure 4-2.

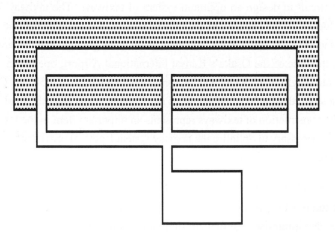

Figure 4-2   System of a parallel taxiway with right angle connections

4.2 HIGH-SPEED EXIT TAXIWAYS

To improve the capacity further, it is necessary to construct one or more rapid exit - high-speed exit taxiways, whose parameters and location need to correspond to the type of operation on the given runway.

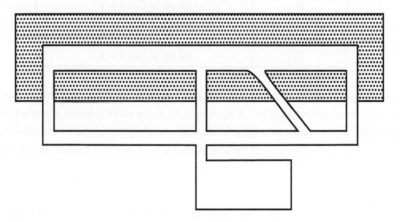

Figure 4-3   System of a parallel taxiway with right angle connections and high-speed exit taxiway

The parameters of high-speed exit taxiways, their dimensions and geometrical characteristics are standardised in the Annex 14 and in the Aerodrome Design Manual. The purpose of the standardisation is to allow the pilot to anticipate the conditions at a less familiar aerodrome in order to avoid prolonging the time of occupation of the runway by failing to make the best use of the turn-off. The parameters of a high-speed exit taxiway, particularly the radius of the turn-off curve, should permit turning off the runway at a speed of up to 93 km.h^{-1} if the runway code number is 3 or 4; and 65 km.h^{-1} if the runway code number is 1 or 2, even if the surface of the runway is wet. The correct location of the beginning of the high-speed exit taxiway from the landing threshold of the runway is important in obtaining optimum use of the facility. It depends on:

✈ speed of the aeroplane crossing the threshold of the runway

✈ deceleration of the aeroplane after its touch-down

✈ initial speed of turn-off.

The aeroplanes have been divided into four groups according to their approach speed overhead the threshold of the runway, as shown in Table 4-1.

Table 4-1   Groups of airplanes according to their speed overhead the threshold

Group	Airplane speed overhead the threshold
A	less than 169 km.h^{-1}
B	169 up to 222 km.h^{-1}
C	223 up to 259 km.h^{-1}
D	260 up to 306 km.h^{-1}

For the purpose of high-speed exit taxiway design, the aeroplanes are assumed to cross the threshold at a speed that is equal to 1.3 V_S, where V_S is the stalling speed of the aeroplane in the landing configuration with an average gross landing weight (85 per cent of the maximum landing mass). Table 4-2 gives examples of ranking of the aeroplanes into groups on the basis of their threshold speed.

Table 4-2   Examples of ranking of the aeroplanes into groups on the basis of their threshold speed

Group A	Group B	Group C	Group D
Convair 240	Convair 600	B-737	B-747
Saab 340	DC-6	B-707	DC-8
DC-3	Fokker F 27	B-727	DC-10
DHC-7		DC-8	IL-62
			L-1011
			Tu-154

The final design of high-speed exit taxiways will depend also on other factors, such as runway slope, aerodrome elevation and reference temperature.

It must be emphasised that the number of high-speed exit taxiways depends on the types and number of aeroplanes intended to utilise the runway during the peak period. The location of the start of a high-speed exit taxiway may be derived by assuming a constant retardation 'a'. The following shall apply at such a movement:

$$\frac{dv}{dt} = \frac{d^2s}{dt^2} = a; \quad v(t) = \int a.dt = a.t + C_1$$

$$s(t) = \int (a.t + C_1).dt = \frac{1}{2}.a.t^2 + C_1.t + C_2$$

If we put the beginning of path s=0 to the point with speed v=0 in the moment of time t=0 (see Figure 4-5) then is $C_1 = C_2 = 0$ and the following expression is generally valid:

$$v(t) = a.t ; \quad s(t) = \frac{1}{2}.a.t^2$$

if the direction to the right from the place of full-stop (i.e. s(0) = 0) is considered to be positive and at the same time the 'acceleration' '**a**' negative. This is illustrated by the Figure 4-4 the meaning of which is expressed by the following:

$$s_1 = \frac{1}{2}.a.t_1^2 ; \quad v_1 = a.t_1 ; \quad t_1 = \frac{v_1}{a}$$

After substitution in s_1 we will get:

$$s_1 = \frac{1}{2}.(a.t_1).t_1 = \frac{1}{2}.v_1.t_1 = \frac{1}{2}.\frac{v_1^2}{a}$$

As an analogy we can derive:

$$s_2 = \frac{1}{2}.\frac{v_2^2}{a}$$

The final expression is:

$$D = s_1 - s_2 = \frac{v_1^2 - v_2^2}{2.a}$$

where is:

a	mean deceleration of the aeroplane after landing $[m.s^{-2}]$
s_1	total braking distance from the touch down (1) to the full–stop point (0) $[m]$
s(t)	distance at time (t) when the beginning of the path is in (0) $[m]$
v(t)	speed at time (t) $[m.s^{-1}]$
v_1	speed over a threshold of the runway $[m.s^{-1}]$
v_2	design speed of turn-off at location (2) $[m.s^{-1}]$
t	time $[s]$
t_1	time required to stop in distance s_1 from the approach speed v_1
t_2	time required to stop in distance s_2 from the speed v_2
D	high-speed exit taxiway distance from a threshold

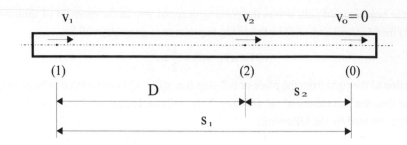

Figure 4-4   High-speed exit taxiway distance from a threshold

The deceleration should not exceed 1.5 m.s^{-2} in order to avoid passenger discomfort. The intersection angle of a high-speed exit taxiway with the runway should not be greater than 45° and less than 25° and preferably should be 30°. It is common practice for the entry to take the form of a spiral, so that the lateral deceleration requirement is not too severe while the aircraft is moving at high speed. This makes it more likely that the exit will be used when the runway friction is reduced. A high-speed exit taxiway should include a sufficient straight portion after the turn-off curve so that the pilot can stop the aeroplane before the next taxiway intersection. Figure 4-5 illustrates an example of a high-speed exit taxiway.

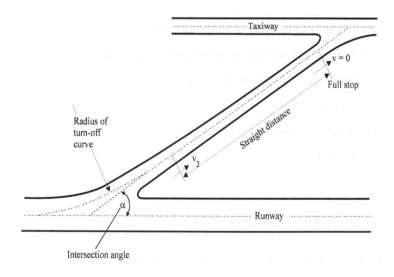

Figure 4-5 Characteristics of high-speed exit taxiway

It is often necessary, where the traffic volume is high, to establish two parallel taxiways to facilitate one way flow, together with holding bays located near the ends of the runway. Holding bays allow the flight controller to bypass aeroplanes and thus optimise the sequence of take-offs depending on the aircraft's speeds, weights and departure routings.

4.3   TAXIWAY CHARACTERISTICS

The minimum safe separation distances between the centre line of a taxiway and the centre line of a runway is defined as a standard in Annex 14. The actual distance depends on the code number of the runway and the category of its approach aids, and it is such that the wingtip of a taxiing aircraft will not encroach into the runway strip. With the introduction of a Code F in Annex 14, the distance might have to be as much as 190 m. Similarly, the minimum safety separation distances are specified between parallel taxiways; these should be 97.5 m to allow unimpeded use by aircraft of up to 80 m wingspan.

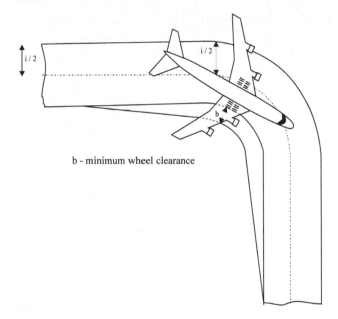

b - minimum wheel clearance

Figure 4-6   Taxiway widening to achieve minimum wheel clearance on curve

The category of approach aid affects the protection which must be given to those electronic aids. ILS signals may suffer interference a stopped or taxiing aircraft particularly if it is a Cat 2 or 3 system.

Depending on the code letter of the runway, minimum margins of safety between 1.5 m to 4.5 m should be provided between the outer main gear wheel edge and the runway edge when the cockpit is over the taxiway centreline. The resulting taxiway width needs to be between 7.5 m to 25 m. Taxiways need to be widened with fillets where they have sharp curves so that the necessary safe separation distance between the outer main gear wheel edge and the runway edge may be maintained when the nosewheel is tracking the centreline, see Figure 4-6. There are commercial programmes available that simulate the tracks of all the wheels for all the main types of aircraft.

Occasional movements of an aeroplane with greater gear track than that for which the taxiway was designed can be accepted by steering the nosegear outwards from the centreline.

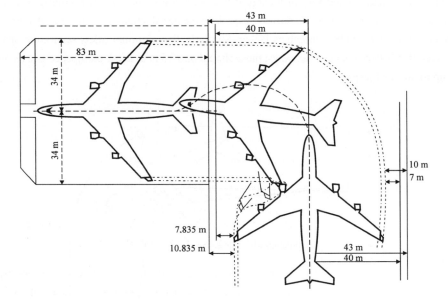

Figure 4-7   **Realigning the front undercarriage guide line outwards from the centreline for B-747-400**

As with runways, taxiways may also be provided with shoulders. Since the speed of aeroplanes moving on a taxiway is quite slow, the main function of the shoulders is to prevent ingestion of foreign objects by an aeroplane's engines. They must also prevent erosion of the surface by engine exhaust gases. They also provide a gradual change from the taxiway bearing strength to that of a taxiway grass strip. The width of a taxiway shoulder is determined by the need to meet these requirements and the aircraft's characteristics. A taxiway shoulder is usually constructed as a light asphalt pavement.

Taxiway strips extend from the edge of a taxiway. Their function is similar to that of a runway strip. For a specified distance from the centre line of a taxiway, the grass surface of the taxiway strip should be well–maintained, without obstacles and the prescribed slopes of the strip should be maintained.

Other physical characteristics of a taxiway, including longitudinal slopes, transverse slopes and a change of the transverse slope of a runway also depend on the code letter of the associated runway. A further requirement is for clear visibility from any point on the taxiway surface to a point on the taxiway 150 m or 300 m away, measured at a height of 1.5 m or 3 m, the distances and heights depending on the code letter of the respective runway.

5

PAVEMENTS

Tony Kazda and Bob Caves

5.1 BACKGROUND

The choice of the kind of pavement depends on the characteristics of the aeroplanes which are intended to use the aerodrome or the respective runway, operational requirements (particularly with respect to reconstruction of the runway) and geological conditions. Requirements for pavement bearing strength, longitudinal and transverse slopes of runways and other movement areas, pavement texture and braking action are all specified by Annex–14 and amplified in the Aerodrome Design Manual, Part 3, Pavements. Operational regulations of individual airport administrations complement these requirements and offer guidance on the occasional excessive loading of pavements.

It is not an objective of this chapter to give instructions how to dimension the pavement structure, which is the construction engineer's responsibility. But nevertheless, a worker in the operational department of an airport should have at least basic knowledge about designing aerodrome pavements so as to be able, based on his understanding, to monitor the pavement condition and manage maintenance and reconstruction. Some characteristics of existing aerodrome pavements differ from those of recent pavements.

The whole aerodrome pavement should comply with three basic requirements:

✈ its bearing strength must be appropriate to the operation of the aeroplanes which are intended to use the aerodrome

✈ it should provide good ride capability of an aeroplane during its movement on the runway by preserving a smooth pavement

✈ it should provide good braking action even on a wet surface.

The first requirement refers to the pavement construction, the second to geometrical characteristics of the surface and the third one to the texture of the pavement surface.

All the three criteria are fundamental and complement one another. It is the only way for a pavement to fulfil the operational requirements. From the operational viewpoint, the most important is the third requirement because it has a direct impact upon the safety and regularity of the aerodrome operation. Thus the requirement for a long-term provision of good longitudinal friction coefficient (f_P) may effect the choice of a pavement construction and surface.

5.2   PAVEMENT TYPES

The choice of a pavement construction type is influenced by many factors.

Table 5-1   Pavement types

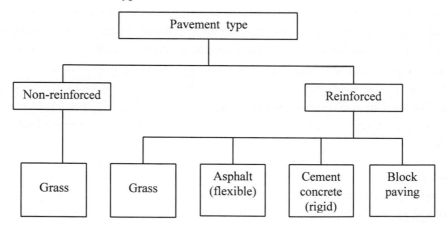

It is necessary to mention in particular the following:

✈ types of aeroplanes from the viewpoint of the maximum point load for which the aerodrome is intended, and other operational requirements

✈ availability and price of suppliers, materials and works

✈ geological conditions.

The basic division of pavements types is given in the Table 5-1.

5.2.1 Non-Reinforced Grass Strips

The non-reinforced grass strips which may be employed in most climatic conditions year-round are suitable only for the lightest types of general aviation aeroplanes. It's suitability depends on appropriate composition and good drainage characteristics of the subsurface. In an ideal case, such an aerodrome would be located on flat land, on a natural layer of gravel covered with approximately a 20 cm layer of topsoil. The surface should be covered with a grass carpet. The type of grass should be chosen so that the roots are sufficiently dense to create a thick and strong carpet in order to reinforce the soil surface. The majority of grasses used are slow growing and therefore do not require frequent mowing. It is unusual to find such an ideal interplay of all factors. The problem is an appropriate water supply. In spring, after a period of melting snow, and in autumn, after longer-lasting rains, the surface may be sodden and its bearing strength reduced. Even if artificial drainage is provided, it is hard to justify the use of non-reinforced grass aerodromes for a business activity because it is difficult to provide the required regularity of operation. In summer, in a period of dry weather and with an intensive operation, the grass carpet will be damaged, sometimes even destroying the roots.

5.2.2 Reinforced Grass Strips

In some cases, it is worth reinforcing a grass runway. During World War II, the Allies used steel grills to build temporary field aerodromes, interconnecting several fields with them. The same method of surface reinforcement is still in use in the army up to the present day, particularly for heavy equipment parking areas.

A modification of that principle consists in laying down netting made from synthetic fibres when the grass carpet is put in place over the upper stratum of topsoil. The netting helps to anchor the roots of individual grass bunches better. At the same time it carries part of the load and takes the friction load from braking of the aeroplanes' tyres.

It some cases it is necessary to increase the bearing strength of a runway but at the same time to maintain its grass surface. Such a pavement is often designed only for temporary use and later must be either returned to its original condition, or used in a different way for agricultural purposes, for example, aerodromes for agricultural operations and for sports activities. A runway may be temporarily reinforced with lime in its bearing layer. Attempts have been made to apply lime even in the upper layer of a pavement. The advantage of that method is its simplicity as well as the possibility of returning the pavement to agriculture by deep ploughing. Calcium is added to the earth either in the form of quicklime-CaO or calcium hydrate- $Ca(OH)_2$. The cementing bonds among the particles of earth are reversible, that is to say, the stabilisation may be broken and made compact again when the bonds reconstitute in the earth. If it is necessary to drain the pavement, it is appropriate to use powder lime with a high percentage of active CaO with an excellent hydration ability.

Occasionally it is required to increase the bearing strength of a runway with a grass surface by constructing the bottom layers in a similar way to those under pavements with an asphalt surface. Then a 15-20 cm layer of soil is spread and a grass carpet is put in place.

5.2.3   Reinforced Pavements with Hard Surface

5.2.3.1 Use of hard surface pavements

In an aerodrome intended for a year-round regular operation of aeroplanes with mass greater than 2 000 kg in European climatic conditions, the use of a hard surface is essential. A choice of the pavement construction is influenced by many factors and the final design is often a compromise. It is possible to cater for even the heaviest aircraft with asphalt runways.

It is appropriate to use cement-concrete- rigid pavements for some movement areas while, in other cases, asphalt-flexible pavements are useful. In general, the types of construction do not differ from constructions used in road building except in the required thickness, which is considerably greater in the aerodrome pavements than in the roads as a consequence of larger point loads.

The construction of a reinforced pavement consists of subgrade, sub-base, bearing course and wearing course.

5.2.3.1.1 Subgrade

The area chosen for an aerodrome should have soil with suitable mechanical properties. The earth in the subgrade should have a sufficient bearing strength but at the same time it must be impermeable and should not change its volume under frost and humidity effects. A soil analysis determines which kinds of soil need to be removed from the aerodrome pavement subgrade, which may be improved by adding different soil, and which soil and minerals in the aerodrome vicinity may be used in its construction.

The choice of type of pavement construction has a considerable effect on the bearing strength of the subgrade. In simple terms, the greater the bearing strength of the subgrade, the relatively thinner and cheaper may be the entire construction of the pavement. The type of pavement is affected also by the availability of suitable building material in the aerodrome vicinity. The choice of type of pavement is decided by an appraisal of life cycle costs, taking into account not only the costs of the pavement construction but also maintenance, and reconstruction of the runway system in a monitored time horizon, including an assessment of the losses due to closing the aerodrome during its repairs and reconstruction.

To declare the bearing strength of aerodrome pavements by means of the ACN-PCN method, it is necessary to know the subgrade bearing strength. For rigid pavements dimensioned according to Westergaard, the bearing strength is expressed by a modulus of reaction of the subgrade 'k'. The modulus of reaction of the subgrade is the contact pressure that is required for pressing a standard loading plate into the subgrade. Either of two methods may be used:

✈ the loading plate is weighted down with a pressure p = 0.07 MPa and then z is determined

✈ the plate is pressed into the depth of 1.27 mm and the contact stress is determined.

The modulus of reaction can then be determined by the relation:

$$k = \frac{P}{z} \; \left[MN.m^{-3} \right]$$

where:

k   modulus of reaction

p   contact stress [N.m^{-2}]

z   loading of the plate [m]

For flexible pavements, the subgrade bearing strength is expressed under the ACN-PCN method by the California bearing ratio (CBR). The California bearing ratio is the ratio of the tested material's bearing strength to the bearing strength of a standard sample of crushed stone, expressed as a percentage. It is determined in a special apparatus by loading tests in which a steel pin with diameter of 5 cm is pressed into the soil.

The dynamic modulus of elasticity (E) is determined by the method of damped impact to simulate running a wheel over the pavement. The effects are produced by a 100 kg weight falling onto rubber pad of prescribed hardness. The deflection of the plate is determined.

The finished and compacted subgrade may be covered with a geotextile which prevents contingent penetration of the subsoil stratum into the pavement construction. The bearing strength of the finished subgrade must be uniform.

5.2.3.1.2 Sub-base

The finished subgrade, which may be protected by a geotextile, is covered by a layer of gravel or crushed stone. That layer is supposed to fulfil a draining and filtration function. It drains the condensation created by temperature variations from the pavement construction and also catches any capillary water. The water is led away by means of catch-drains to collectors.

5.2.3.1.3 Bearing course

The role of a bearing course is to receive and distribute the pressures from the aeroplane undercarriage to an appropriately large area of sub-base and subgrade. The bearing course usually has several layers. The thickness and composition of individual stratums depend on the subgrade bearing strength and on the construction of the wearing course. In order that the pavement may be designed economically, the upper layer should always have greater bearing strength than the layer under it. As a bearing course, it is possible to use clay or argilliferous stabilisation, cement stabilisation (gravel with fine fractions admixed with 8 to 10 % of cement), macadam or gravel.

5.2.3.2 Flexible (asphalt) pavements

In the late 1970's the use of asphalt pavements became popular. They are now used for types of movement areas and loads which in the past only cement-concrete pavements were used. At comparable costs, the asphalt pavements have several advantages. The construction of asphalt pavements is less demanding

It is simpler and less expensive to carry out repairs and renovation on asphalt pavements. The reconstruction of asphalt pavements may be carried out even without interrupting operations if the works are performed by night. Asphalt pavements also withstand winter maintenance better when chemical de-icing materials are used. The asphalt pavements have the further advantage that their surface is even and without joints.

On the other hand, the asphalt pavements are less used in military aerodromes due to their reduced resistance against the impact of hot exhaust gasses from military jet engines. Their resistance against spilt fuel is also lower.

The relatively lower bearing strength of asphalt pavements is due to the different manner of transmitting the load. With asphalt pavements, the load is transmitted by an interaction of individual material particles under the effect of physical bonds of asphalt. The bearing strength is limited by the load that causes a permanent deformation of the flexible asphalt layer.

The upper part of the asphalt pavement is usually composed of two asphalt layers which have different functions. A supporting asphalt layer containing coarse gravel fractions is put in place on the bearing course. Its role is to transmit the load on the bearing layers. Its thickness depends on the required resultant bearing strength, or when reconstruction is carried out, also on the condition and profile of the pavement underneath. Depending on the total thickness, which is usually within the range from 10 to 40 cm, the asphalt layer may be spread several times by a finisher in order that the required compactness can be obtained.

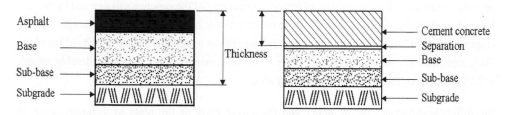

Figure 5-1   Typical construction of asphalt and cement concrete pavement

The upper wearing layer contains finer fractions of quality aggregate. The function of the wearing layer is to resist friction forces that are created by braking on landing, and during a rejected take-off and turning of the aeroplane. The pavement surface roughness is designed to ensure appropriate braking action. In order that the wearing layer can resist these forces, it should be at least 4 cm thick. Its second function is to create an integral impermeable surface. It must perfectly seal the entire construction of the pavement. If water penetrates into the subgrade, it would gradually erode, lose bearing strength, and subsequently lead to rupture of the bearing courses. The wearing course also transmits the load to the layers underneath.

Besides other things, it is important to monitor the temperature of the asphalt composition during the construction. Maintenance of the prescribed temperature is a precondition of achieving the required compactness of the strata and the bonding of the strips. It is often at badly treated joints where ruptures appear.

When considering the bearing strength of the asphalt pavement, one of the decisive factors is the overall thickness of the pavement 'h' including the bearing course.

5.2.3.3 Rigid (cement-concrete) pavements

The main advantage of rigid pavements is their higher bearing strength which is derived, among other things, from different transmission of the load. The upper layer, a cement-concrete plate,]rests on a semi-flexible subgrade. The rigidity of the cement-concrete plate depends on the quality of the mixture. The plate transmits the load to a considerably greater area. Inasmuch as the rigid wearing courses are not flexible enough to be able to follow even the slightest deformations of the subgrade, they only may be designed on a quality subgrade with an even bearing strength.

Another advantage is a longer design life of the cement-concrete plate. The design life of a rigid pavement on an appropriate subgrade and with proper maintenance may be 20 to 30 years.

A cement-concrete plate is usually made from plain concrete 20 to 30 cm thick. The thickness of the plate is limited by the possibilities of regular compacting of the concrete mixture and by the fact that with an increasing thickness of the plate the magnitude of the internal stress increases due to temperature variations causing differential expansion of the upper and the lower parts of the plate. The temperature gradient is about 0.5 °C per 1 cm. As a matter of fact, it is considerably greater in the upper part of the plate up to a depth of about 3 cm.

The different expansion of the upper and the lower parts of the plate as a consequence of temperature variations manifests itself in a tendency to deform the cement-concrete plate into a convex or concave shape. The internal forces reduce the bearing strength of the plate in comparison with its normal condition and are taken into account during design. A sliding asphalt intermediate layer can be put in place on the subgrade, which allows the dilation of the cement-concrete plate.

The cement-concrete plate is laid by means of a finisher. On aerodromes, finishers are used that are capable of putting a 15 m wide or smaller plate in place. The treated and compacted concrete mixture behind the finisher has a smooth surface. Therefore the flaccid concrete mixture should be roughened generally with plastic brushes which are pulled perpendicularly to the longitudinal centre line of the runway.

In order to prevent the cement-concrete plate from developing irregular cracking under the effects of the internal stress at differential expansion of the upper and the lower parts of the plate, the concrete-cement strip should be divided by contraction and expansion joints into individual plates.

The contraction joints (Figure 5-2) with a width of approximately 5 mm are cut in the hard-set concrete mixture to approximately 1/4 of the plate thickness. In the subsequent setting, as a consequence of the internal stresses, it will break under the incised joint. The contact surfaces are uneven with irregularities which anchor the neighbouring plates and provide them with a support. Thus the concrete strip is divided by the contraction joints into plates from 3 to 9 m long, depending on the pavement width, its thickness and quality of the concrete mixture.

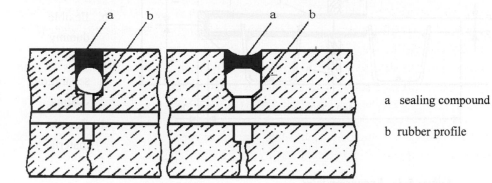

a   sealing compound

b   rubber profile

Figure 5-2   *Contraction joint*

The expansion joints divide the cement-concrete plate through its entire thickness and facilitate shortening and expanding of the pavement. They are located at distances of approximately 40 m. The neighbouring plates on both sides of the expansion joint should be anchored. The manner of anchoring is illustrated in Figure 5-3.

The contraction and expansion joints should be perfectly sealed. If water penetrates and erodes the bearing courses and the subgrade, the concrete plate would lose its support and would crack under load. The subsequent repair would be very expensive. The joints are sealed with different types of rubber and then by a filler. The filler should perfectly adhere to the joint walls and must be permanently flexible. In winter during freezing it must not crack, and in summer with high temperatures it must not flow. The base of fillers consists of asphalt with additives of rubber and plasticizing agents.

The quality of the concrete mixture must be controlled during the construction, samples being tested. The concrete will reach the required rigidity after 28 days. The pavement is dimensioned by the Westergaard method.

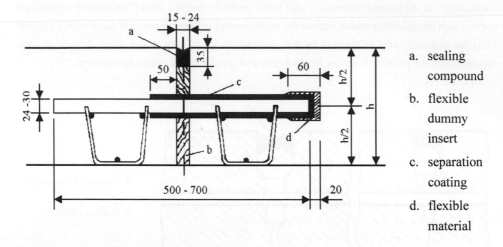

a. sealing compound

b. flexible dummy insert

c. separation coating

d. flexible material

Figure 5-3   Expansion joint

Aprons are critical points of the movement areas. There the aeroplane has its maximum mass. Similarly, the taxiways are highly stressed by a slowly moving aeroplane with the engines running. It is often thought that the runway is stressed maximally in the touch down area. In fact, at touchdown, the aeroplane has still a lift from the wings so that the load seldom exceeds 40 per cent of the aeroplane maximum mass.

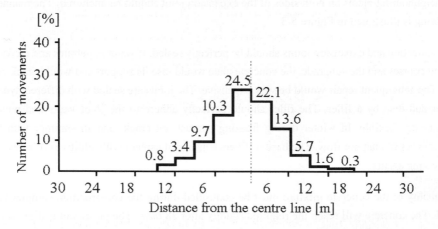

Figure 5-4   Probability of transverse distribution of the aeroplane movements on the RWY

Taking into account the probability of transverse distribution of the aeroplane movements on the runway, it is possible to dimension the maximum load only in the central part of the pavement (Figure 5-4).

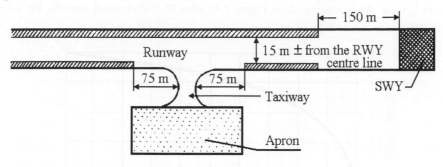

Figure 5-5a   Optimizing pavement design – RWY equipped with a parallel TWY
Source: Aerodrome Design Manual, Part 3, Pavements, ICAO Doc 9157-AN/901

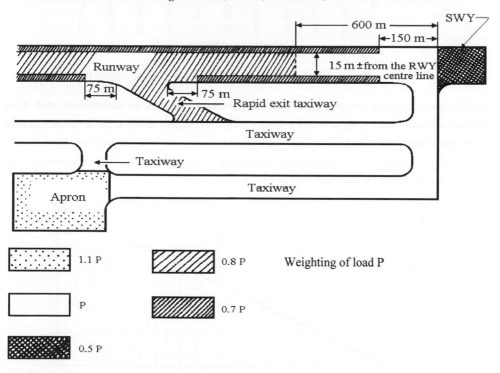

Figure 5-5b   Optimizing pavement design – RWY not equipped with a parallel TWY
Source: Aerodrome Design Manual, Part 3, Pavements, ICAO Doc 9157-AN/901

As with other types of construction, the aerodrome pavement has a fatigue life. The pavement is designed, for example, for 10 000 movements of a given limit load. The rigidity declines as the pavement is used. As is apparent from Figure 5-6, after 10 000 repeated stresses, the tensile strength declines to almost half-value.

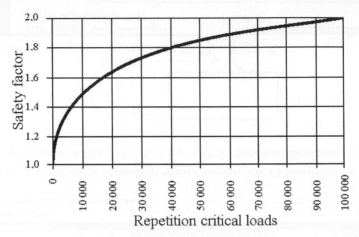

Figure 5-6   Effect of critical loads repetition on pavement tensile strength

On the other hand, the strength of the concrete increases with time as the concrete ages. That property of the concrete may to a certain degree reduce the impact of the stress from the repeated loads, particularly if the loads from larger aircraft only occur several years after completion of its construction. The degree of strengthening is dependent on the composition of the concrete mixture and its quality, the rate being illustrated in Figure 5-7.

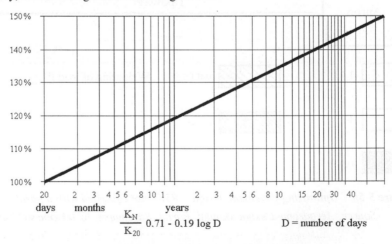

$$\frac{K_N}{K_{20}} \quad 0.71 - 0.19 \log D \qquad D = \text{number of days}$$

Figure 5-7   Increase of concrete strength with time

The resultant characteristics of the cement-concrete plate, when evaluating the construction of the pavement in relation to the load characteristics of the aeroplane undercarriage, is expressed by the semi-diameter of its relative rigidity 'l' by the relation:

$$1 = \sqrt[4]{\frac{D}{k}} = \sqrt[4]{\frac{E.h^3}{12(1-\mu)k}} \qquad\qquad [m]$$

where:

D   bending rigidity of the plate $D = E.h^3/12(1-\mu^2)$

E   modulus of concrete elasticity in tension and compression $[N.m^{-2}]$

h   thickness of the cement-concrete plate $[m]$

μ   Poisson constant, usually $\mu = 0.15$ [non-dimensional]

k   modulus of cubic compressibility or 'modulus of subgrade reaction' (5.2.3.1.) $[N.m^{-3}]$.

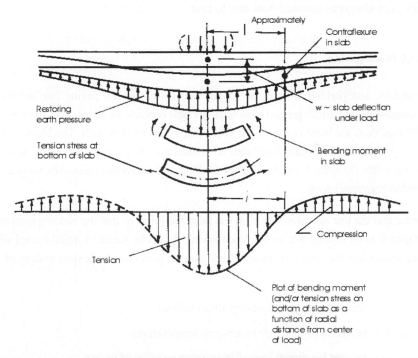

Figure 5-8   ***Physical meaning of Westergaards 'radius of relative stiffness 'l'***
Source: Aerodrome Design Manual, Part 3 Pavements, ICAO Doc 9157-AN/901

The semi-diameter of relative rigidity is graphically illustrated in Figure 5-8. The pavement bearing strength is influenced significantly by:

✈ quality of the concrete expressed in the modulus of concrete elasticity **E**

✈ thickness of the cement-concrete plate **h**

✈ quality of the subgrade characterised by 'modulus of subgrade reaction' **k.**

In practice the real bearing strength differs from theoretical values and is influenced by many factors. Therefore the real bearing strength in practice is determined by loading tests currently used in the building industry.

5.2.3.4 Combined pavements

There are often cases in practice where reconstruction of a rigid pavement is performed by laying a new cover of several layers of asphalt Reconstruction has both advantages and disadvantages which must always be assessed from case to case.

5.2.3.5 Block paving

For aprons, and parking areas in particular, the use of special paving has become normal. Rectangular paving bricks pressed from high-strength concrete of dimensions 100 mm x 200 mm by 80 mm thick are most common, but other shapes are used as well. The blocks are laid into coarse-grained and sharp sand. Finer sand is vibrated into the joints by which the blocks are fixed one to another. By mutual 'interlocking' of the blocks, the pavement transmits vertical loads on a considerably greater area.

In order that the fine sand does not blow out from the joints or that the bedding sand layer is not damaged if an oil spillage occurs, the block paving should be sealed. A liquid is used which, after being spread into the joints, polymerises and creates permanently flexible sealing of the block paving.

The main advantages of the block paving are as follows:

✈ they may be laid even in winter in sub-zero temperatures

✈ they may be used by aircraft immediately on completion of laying

✈ maintenance of pavements is speedy and simple.

5.3   PAVEMENT STRENGTH

5.3.1   Pavements-Aircraft Loads

There is a general rule that the bearing strength of runways, taxiways and aprons should be capable of withstanding the maximum load derived from the aeroplane which the aerodrome may expect to serve. Such an aeroplane is called a critical aeroplane and the derived load is the critical load.

It is difficult to determine the bearing strength of non-reinforced grass areas. The bearing strength is dependent on the quality of the grass surface and the bearing strength of the soil. The bearing strength of the soil may vary very quickly, depending on the humidity. In practice, the capability of a grass area may be assessed visually, or even better, by repeated rides through the area in a car. In this way it is possible to discover places with locally lower bearing strength. A greater measure of objectivity may be provided by means of a metal stick, ended with a scaled taper. According to the depth the taper penetrates into the soil when the stick is dropped from a specified height, an approximate bearing strength of the grass surface may be determined.

The load an aeroplane applies to a pavement depends not only on the total mass of the aeroplane but also on other factors, in particular on:

✈ type of undercarriage

✈ number of wheels on the main undercarriage leg

✈ geometric configuration of the wheels in the undercarriage

✈ tire pressure.

The critical loading is generated by the main undercarriage leg. In the simplest case, the main leg has only one wheel. Then it is possible to determine the load magnitude from the size of the contact area of the tire with the pavement surface in dependence on the aeroplane mass which is applied to one undercarriage leg and on tire pressure:

$$A = Q/P_0$$

where:

A   contact area of tire $[m^2]$

Q   load transmitted by the main undercarriage leg $[N]$

P_0 tire pressure p $[Pa]$ multiplied by a contraction coefficient of the tire m which expresses the resistance generated by its rigidity (values within the range from 1.03 to 1.1) $P_0=pm$

The contact area has approximately the shape of an ellipse.

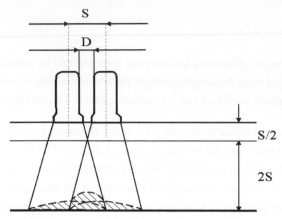

A more complicated event occurs if the main undercarriage leg has two, four, or more wheels. Then the loading effects generated by individual wheels must be partially summed, depending on the distance of one wheel from another, as well as on the pavement properties. The manner of summing the loads is schematically illustrated in the Figure5-9.

Figure 5-9   Scheme of load distribution from dual wheel undercarriage in the pavement layers

At the same time it is apparent from Figure 5-9 that the resultant load is dependent on the pavement construction thickness. With a more detailed analysis it is possible to determine also the dependence on the quality of individual courses of the pavement.

The effect of a multi-wheel undercarriage is expressed by the Equivalent Single Isolated Wheel Load. This is the theoretical load generated by a single wheel which generates the same effects in the pavement as a real multi-wheel undercarriage. The tire pressure is the same in the two cases.

5.3.2   Pavement Strength Reporting

The information on the bearing strength of an aerodrome pavement is necessary:

✈ to ensure the integrity of the pavement to its optimum design life

✈ to determine the types of aircraft that may use the pavement and their maximum operational masses

✈ to design the aircraft undercarriages so that it may be used in the majority of the current aerodromes.

Annex 14 prescribes the consistent manner in which the bearing strength of movement areas is to be reported, thus encouraging its world-wide standardisation. The new method includes two procedures.

The first procedure is related to the pavements intended for aircraft up to maximum mass of 5 700 kg. The bearing strength is determined by maximum allowable mass and maximum tire pressure, see Table 5-2.

Table 5-2   Reporting of the bearing strength of a pavement intended for aircraft of apron mass equal or less than 5 700 kg

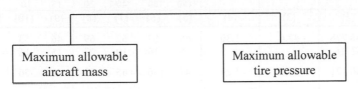

| Maximum allowable aircraft mass | Maximum allowable tire pressure |

The second procedure is called ACN-PCN (Aircraft Classification Number-Pavement Classification Number).

ACN is the number indicating the relative effect of an aeroplane on the pavement resting on any of four types of subgrade of standard quality.

PCN is the number indicating the bearing strength of a pavement with an unlimited number of aircraft movements.

The ACN-PCN method expresses the bearing strength of a pavement in whole numbers between zero and an unspecified maximum. The same scale is used for expressing the load characteristics of aeroplanes. A single wheel with a weight of 500 kg and tire pressure of 1.25 MPa represents a load of 1 ACN. The bearing strength of a pavement with a bearing strength that just corresponds to that load registers 1 PCN. In other words, a PCN for an unlimited number of movements should be equal or higher that the ACN.

The ACN-PCN method is a very simple method of pavement strength reporting from an operational viewpoint. It does, however, require the ACN of individual aeroplanes to be published. The ACN-PCN is not intended for designing pavements, neither does it determine a method by which the bearing strength of a pavement should be specified. On the contrary, an airport operator may use an arbitrary method for PCN determination.

The ACN of an aeroplane is determined by means of two mathematical models. One of them is applicable for rigid, the other for flexible pavements.

Table 5-3   ACNs for several aircraft types

Aircraft type	All-up mass [kg]	Load on one main gear leg [%]	Tire pressure [MPa]	ACN for rigid pavment subgrades [MN/m³]				ACN for Flexibile pavement subgrades [CBR]			
				High 150	Medium 80	Low 40	Ultra low 20	High 15	Medium 10	Low 6	Very low 3
(1)	(2)	(3)	(4)	(5)	(6)	(7)	(8)	(9)	(10)	(11)	(12)
DC-8-55	148 778	47.0	1.30	45	53	62	69	46	53	63	78
	62716			15	16	19	22	15	16	18	24
DC-8-61/71	160 121	46.5	1.29	47	56	65	73	49	56	67	83
	65 025			15	16	19	22	16	16	18	24
DC-8-62/72	162 386	47.6	1.34	50	60	69	78	52	59	71	87
	72 002			17	19	23	26	18	19	22	29
DC-9-15	41 504	46.2	0.90	23	25	26	28	21	22	26	28
	22 300			11	12	13	14	10	11	12	14
DC-9-21	45 813	47.15	0.98	27	29	30	32	24	26	29	32
	23 879			12	13	14	15	11	12	13	15
DC-9-2	49 442	46.2	1.05	29	31	33	34	26	28	31	34
	25 789			14	15	15	16	12	13	14	16
DC-9-41	52 163	46.65	1.10	32	34	35	37	28	30	33	37
	27 821			15	16	17	18	13	14	15	18
DC-9-51	55 338	47	1.17	35	37	39	40	31	32	36	39
	29 336			17	17	18	19	15	15	16	19
DC-10-10	196 406	47.15	1.28	45	52	63	73	52	57	68	93
	108 940			23	25	28	33	26	27	30	38
DC-10-10	200 942	46.85	1.31	46	54	64	75	54	58	69	96
	105 279			22	24	27	31	24	25	28	36
DC-10-15	207 746	46.65	1.34	48	56	67	74	55	61	72	100
	105 279			22	24	27	31	24	25	28	36
DHC 7 DASH 7	19 867	46.75	0.74	11	12	13	13	10	11	12	14
	11 793			6	6	7	7	5	6	6	8
FOKKE R	19 777	47.5	0.54	10	11	12	12	8	10	12	13
27 Mk 500	11 879			5	6	6	7	4	5	6	7

A computer programme was elaborated by the Portland Cement Association of the USA and was provided to ICAO to determine the ACN of the majority of aeroplanes in service. The ACN values are given in an Aerodrome Design Manual, Part 3. Similarly, in the case of new types of aeroplanes, ICAO may determine ACN for the given type of aeroplane based on the data provided.

For calculation of the actual ACN for arbitrary mass between the maximum and empty mass of an aeroplane, a linear function is applied. An overview of the ACN-PCN method of data coding is given in the Table 5-4.

The ACN-PCN method publishes the required data in the form of a code. The data are given for two different types of pavements, rigid R, and flexible F, for eight standard categories of subgrade (four values of k for rigid pavements and four values of CBR for flexible ones). Then the system uses four categories of tire pressure. The code contains also information on the method of determining PCN. When PCN is determined on the basis of operational experience, according to the ACN of the critical aeroplane, a U code (Using Aircraft Experience) is used. If PCN is determined by any of the more sophisticated methods used in civil engineering, a T code (Technical Evaluation) is used. Table 5-3 gives data of several types of aeroplanes as they are published by ICAO.

Table 5-4   Reporting of the bearing strength of a pavement intended for aircraft of apron mass greater than 5 700 kg

PNC Pavement clasification number	Code	Pavement type	Code	Subgrade strength category	Code	Tire pressure	Code	Evaluation method
Bearing strength without limitation of a/c movements	R	Rigid	A	High $K=150MN.m^{-3}$ $CBR = 15\%$	W	High (no pressure limit)	T	Technical evaluation
	F	Flexible			X	Medium up to 1.50MPa	U	Using aircraft experience
			B	Medium $K=80 MN.m^{-3}$ $CBR = 10\%$				
					Y	Low up to 1.00 MPa		
			C	Low $K=40 MN.m^{-3}$ $CBR = 6\%$	Z	Very low up to 0.50 MPa		
			D	Ultra low $K=20 MN.m^{-3}$ $CBR = 3\%$				

5.3.3   Overload Operations

Under certain conditions, the pavement can bear even greater load than that calculated. When it is overloaded, it does not break up suddenly. Loads which are larger than the defined loads shorten the design life of the pavement, whilst smaller loads extend it. Overloading of pavements can result from loads too large, or from a substantially increased number of movements. With the exception of massive overloading, pavements are not subject to a particular limiting load when under static loading. Their behaviour is such that a pavement can sustain a definable load for an expected number of repetitions during its design life. Occasional minor overloading only results in limited loss in pavement life expectancy. Therefore, for such cases, the following criteria are suggested:

For flexible pavements-occasional movements by aircraft with ACN not exceeding 10 per cent above the reported PCN should not adversely affect the pavement.

For rigid or composite pavements, in which a cement-concrete plate provides the crucial transmission of the load, occasional movements by aircraft with ACN not exceeding 5 per cent above the reported PCN should not adversely affect the pavement.

If the pavement structure is unknown, the 5 per cent limitation should apply. The annual number of overload movements should not exceed 5 per cent of the total annual aircraft movements. Such overload should not be permitted on pavements that are damaged or exhibit other signs of distress, after strong freezing, as well as in cases when the bearing strength of the pavement or its subgrade could be weakened by water.

5.4   RUNWAY SURFACE

5.4.1   Runway Surface Quality Requirements

The runway surface quality requirements have gradually increased with implementation of ever heavier jet aeroplanes and the increasing speed of aeroplane movements on runways. The runway surface must not show unevenness sufficient to cause a loss of braking action, it must provide good braking action even when the surface is wet and should be cleaned of any foreign objects which might damage the engine. Annex 14 describes in detail the standards which specify the particular physical characteristics of the surface.

The smoothness of the runway surface is defined by several standards, which complement one another. The basic requirement for cement-concrete pavements is a test with a 4 m long board under which no unevenness larger than 5 mm may appear with the exception of drainage channels. In order that an adequate smoothness of the surface can be ensured, it is necessary to choose the appropriate types of inset lights, back-inlet gulleys and the like.

The manual detection of unevenness by means of a 4 m long board is lengthy, time consuming and little accurate. An 'automatic' board, or Viagraf, an equipment with automatic registration of unevenness, is often used instead. However, it can only register an unevenness up to a length of approximately 2 m.

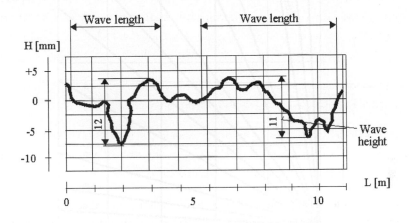

Figure 5-10   Surface unevenness diagram

5.4.2   Methods of the Runway Surface Unevenness Assessment by the DMS Method

At higher speeds of aeroplane movement on the runway, unevenness with small wave lengths loses its effect, and unevenness with wave lengths of the order of tens of meters acquires importance. That unevenness may be detected by means of a precise survey which is carried out after construction or reconstruction of the runway. Thus the supplier generally proves that the construction of the pavement complies with the respective standards within the permitted tolerances. However, once the aerodrome is in operation, it is practically impossible to make exact surveys of the runway. It is too expensive and time-consuming.

From the operational viewpoint, it is essential to know whether some unevenness complies or not with a certain constructional standard. For example, transversal grooving of the runway represents unevenness that does not conform to the standard 4 m board. However, not only does the grooving not impede the aeroplane's operation but it generally increases the coefficient of the pavement longitudinal friction and improves the braking action.

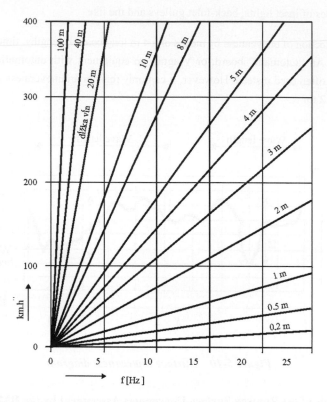

Figure 5-11*   *Relationship of oscillation frequency of the undercarriage wheel and speed of movement of the aeroplane on the runway

Therefore the conclusive criterion should be the effect of the pavement on the aeroplane that is moving on it at a certain speed. The condition of the pavement will have been gradually deteriorating since commissioning. Differential settlement of the subgrade, the effect of the operation of aeroplanes, breakdowns of the pavement and their repairs step-by-step worsen the pavement unevenness. Small deviations from the specified tolerances have no substantial effect on the operation, but above some critical level, unevenness of the runway surface will endanger the safety of the aeroplane during take-off or landing. Unevenness up to approximately 3 cm in the distance of 50 m is considered acceptable. However it must always be judged in relation to the type of operation on the runway, and in particular, to the speed of the aeroplane on the runway. The frequency of oscillation of the undercarriage wheel depends on the speed of movement of the aeroplane on the runway at the given length of unevenness. It can be expressed by:

$$f = v/\lambda \ [\ s^{-1}]$$

where:

f   frequency of oscillation of the undercarriage wheel

v   speed of movement of the aeroplane on the runway

λ   length of the wave in the pavement

An oscillation with a frequency of more than 20 Hz is not significant because it is damped by dampers. Similarly, an oscillation less than 1 Hz is not dangerous because it develops only small vertical acceleration. The oscillation frequencies between 2 and 10 Hz are particularly dangerous. The measurement and evaluation of unevenness of the runway surface can be performed by means of a truck pulled behind a car and an accelerometer located on the truck which reacts to the surface unevenness. Permanent contact of the system with the surface must be ensured. The method is based on the use of the theory of accidental functions. The unevenness of a length of wave up of 100 m may be measured.

The chosen speed of movement of the measuring system may be from 5 to 40 $m.s^{-1}$, (i.e. from 18 to 144 $km.h^{-1}$). The equipment measures the vertical acceleration and provides information on the spectrum of the wave lengths (from 2.7 to 100 m). The measurement should be carried out on a portion 500 m long as a minimum which is homogeneous from the viewpoint of the kind of covering, treatment and the surface condition.

The basic statistical characteristic of the vertical unevenness is its scatter. Information on the intensity with which it occurs within the entire scatter of unevenness of particular wave lengths is given by the power spectral density of unevenness.

When an assessment is made of the dynamic response to the aeroplane moving on the runway, the following three factors should be monitored:

✈ impact of unevenness on the crew

✈ impact of unevenness on the passengers

✈ impact of unevenness on the aeroplane construction from the viewpoint of the strength and fatigue stress and from the viewpoint of the aircraft's controllability and the safety of its movement on the runway.

Other assessment criteria of the DMS method are the parameter C and zone classification included in the standard. The parameters determined by the DMS method are not mutually compatible with the results obtained by other methods.

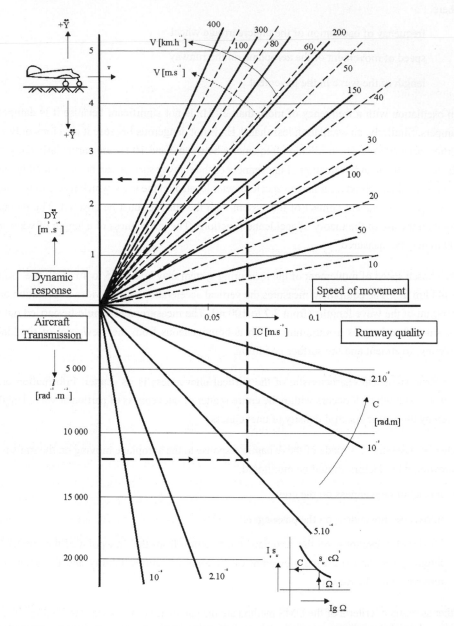

Figure 5-12   Dynamic response of aircraft vertical acceleration on different pavement quality

where: DŸ   *square of acceleration* $[m^2 s^{-4}]$

IC   *variation of acceleration* $[ms^{-3}]$

5.4.3 Pavement Texture

The pavement surface should have specific antiskid properties, the properties being given by its micro-texture and macro-texture.

If the pavement is clean and dry, it generally has also good characteristics of the coefficient of longitudinal friction whatever the kind of surface (asphalt or concrete-cement pavement), or its drainage characteristics (pavement slope). In addition, the coefficient of friction of a dry pavement only changes very slightly with an increase of the aeroplane speed.

Whenever water in any quantity is present on a runway, the situation changes considerably. The coefficient of friction then depends substantially on the quality of pavement surface in terms of the surface texture and drainage characteristics of the pavement, particularly any tendency to pooling of water. At high aircraft speeds the coefficient of friction may drop considerably as a consequence of aquaplaning. Even so, the pavement must have a good coefficient of friction even when wet in order that the safety of air operation is not endangered. The coefficient of longitudinal friction of the pavement depends on the following:

✈ macro-texture created by the binding of individual gravel fractions in the mixture, this controlling contact between the tire and the pavement surface

✈ micro-texture characterised by the surface roughness of the grains of aggregates or mortar. Micro-texture is important for the so called 'dry contact' of a tire on a wet pavement.

The macro-texture of a pavement is supposed to ensure a good water drainage from the area in contact with the tire. Therefore it is particularly important at high aircraft speeds t. Macro-texture of a common road pavement is smoother, that of an aerodrome pavement is more open. This is due to different drainage characteristics of car and aeroplane tires, and the higher speeds that are reached by aeroplanes on a runway. A correct macro-texture may be obtained in a concrete surface as well as in an asphalt pavement surface. In cement-concrete pavements it is obtained by transversal brushing of the setting concrete. In asphalt pavements it is achieved by using correct fractions of the aggregates in the wearing layer and the correct procedure in laying the pavement.

The micro-texture of a pavement is formed by fine, but sharp irregularities on the grain surface which are capable of penetrating through the thin viscous film of water between the tire and the pavement. It provides a contact with the tire even at higher speeds. The micro-texture is created by the 'micro-roughness' of the individual surface particles. It is difficult to discern with the naked eye, but is discernible by touch. Therefore only quality, hard and rough aggregates should be used in the pavement wearing layer. The effect of the macro-texture and micro-texture on the coefficient of longitudinal friction depends on the aeroplane speed as illustrated in the Figure 5-13.

The greatest problem of the micro-texture is the fact that it may change in a relatively short time. A typical example is rubberising of the runway in the touch-down zone. The rubber deposits may cover the micro-texture without any change of macro-texture. As a result there is a substantial reduction of the coefficient of longitudinal friction when the pavement is wet.

While the macro-texture is characterised by its depth hP, no objective criterion has been specified to determine the micro-texture. Based on experience it may be stated that the individual particles must be rough and have sharp edges. It is indirectly assessed by the coefficient of longitudinal friction.

At the time of construction the roughness should be optimum. The antiskid properties of the pavement surface can be determined by two methods:

✈ determining the depth of the macro-texture by means of sand

✈ determining the coefficient of longitudinal friction by means of a dynamometer.

The sand method is generally used during commissioning, while the dynamometer method is used for assessing the operational capability of the surface.

In the sand method, an average depth of the potholes is calculated from the surface of a circle in which all the unevenness is filled with an exactly determined quantity of sand. For the trials, 90 in their entirety, the sand of a specified granularity is used on the whole runway, dried at a temperature of 105°C and sealed in PVC bags each with 0.25 dm^3. The depth of macro-texture h_P for 90 % of all measurements should comply with the condition $h_P =/> 0.8$ mm.

In the dynamometer method the coefficient of longitudinal friction is calculated from the force required for pulling the braked measuring wheel on a wet pavement surface. The friction ought to be determined:

✈ on a periodical basis on a runway where the code number is 3 or 4 for calibration

✈ on a runway which is slippery when wet

✈ on a runway which has bad drainage characteristics.

The calibration measurement should be performed in good meteorological conditions on a dry and clean runway. In order to ensure homogenous conditions, the measuring device must be provided with self-wetting equipment. Thus the measurement is effectively carried out on a wet surface. The runway surface is wetted before measuring, and the quantity of water should be the same at all speeds. The measuring is performed with a bracked wheel (100 per cent slip) at the speeds of 20, 40, 60, 80 and 100 km.h^{-1}, sometimes up to 135 km.h^{-1}.

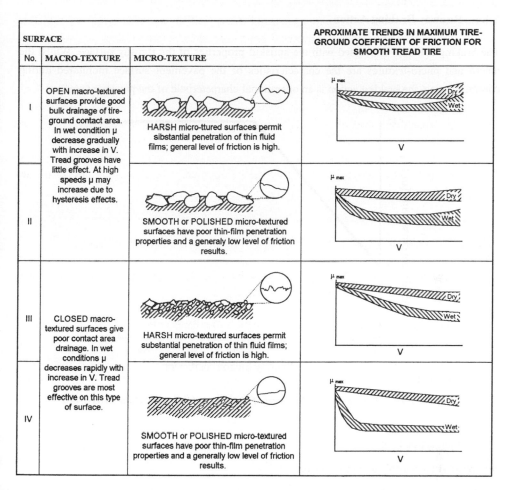

Figure 5-13   Effect of surface texture on tire-surface coefficient of friction

The intervals of measurements should be determined as a function of the type of aircraft, the number of movements, the type of pavement and its maintenance, climatic conditions, etc.

If there is any suspicion that the braking characteristics may be impaired because the runway has a transverse slope or other inconvenient drainage characteristics, an additional measurement should be carried out when it is actually raining.

The measurements are performed with special vehicles, e.g. Saab Friction Tester, or by dynamometric trailers pulled behind a car, e.g. μ-meter. These each use different principles of measurement of the coefficient of longitudinal friction.

5.4.4   Runway Braking Action

The braking action is connected with the antiskid properties of the runway surface. Whilst the macro and micro-textures are the characteristics of the pavement surface monitored during commissioning, the braking action is an operational characteristic of the pavement surface.

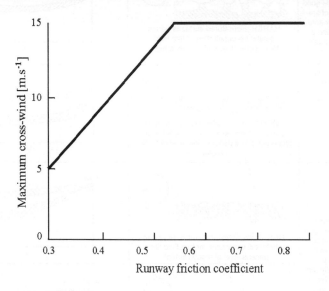

Figure 5-14   Relationship of runway friction coefficient and the crosswind

The braking action is influenced by:

✈ speed of the aeroplane on the runway

✈ design of the aeroplane

✈ design of the brakes

✈ type of automated antilocking device

✈ dimensions and number of tires

✈ tire tread pattern and wear

✈ aeroplane mass

✈ meteorological conditions

✈ as well as the coefficient of longitudinal friction on the runway surface.

Whilst the friction coefficient of a certain pavement decreases slowly over several years, the braking action may change from minute to minute. Determination of the braking action is a part of the control of the runway operational capability which should be carried out by the airport operator. Besides the frequent periodical inspections, additional checks should be performed, e.g. at each change of climatic condition. The results should be reported to the Air Traffic Controller. Since the braking action is affected by many factors, it is important to understand what is happening in the contact surface between the tire and the pavement.

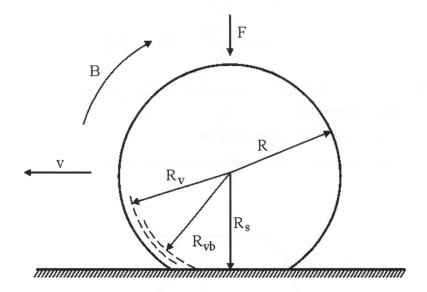

Figure 5-15   Relative slippage of the wheel – parameters

In the contact between the tire and the runway surface, a footprint is created. Its size depends on the radial load and the tire dimensions. When the tire is loaded only by a radial force, the elementary forces in the contact area are in static equilibrium. If a force other than the radial one begins to exert its influence on the wheel, i.e. if the wheel begins to transmit peripheral or lateral force due to braking or cross wind, the distribution of the elementary forces in the tire footprint will change. The resultant force in the contact area of the tire is called a force of adhesion. The force bond between the tire and the runway surface is determined by the adhesion. Therefore it is important from the viewpoint of the air operation safety to create good adhesion conditions.

Another important characteristic is the wheel slip. An absolutely rigid wheel would roll on the support on a radius R (Figure 5-15.). If we load a real wheel with a tire, it will deform. The radius from the centre of the wheel to the footprint will diminish to a static radius R_S. When the wheel is rolled on a pavement, a fictitious radius called a rolling radius R_V may be considered. It is the radius of a perfectly rigid wheel whose peripheral velocity and rolling velocity are the same as the peripheral velocity and rolling velocity of a real tire. Here $R > R_V > R_S$ applies. From the comparison R and R_V, the so called relative slippage of the wheel may be determined.

If we load the wheel with an additional tangential braking force, it will deform further, and the wheel will roll on the radius R_{VB}, generating another tangential slippage. The following applies for the assessment of a relative slippage:

$$S = \frac{R_v - R_{vb}}{R_v} \cdot 100 \quad [\%]$$

where:

S   relative slippage [m]

R_V rolling radius of the tire in free rolling [m]

R_{VB} rolling radius of the tire when loaded with braking torque [m].

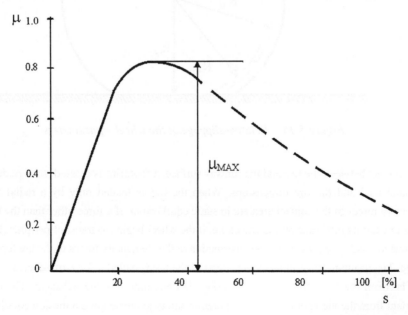

Figure 5-16   Relationship between percentage of slip and friction coefficient

It must be emphasised that the relation applies for a relative slippage, that is to say while the wheel is turning rather than being locked up.

In free rolling the relative slippage is zero. The relationship between the force of adhesion and the slip determines the most efficient manner of braking. The relationship is of the general form shown in Figure 5-16

The maximum adhesion occurs with modern aeroplanes between 15 and 30 % of slippage, and the brake antilocking systems are generally adjusted to that value. The shape of the slip characteristic depends on the runway surface condition. If the runway is covered with water, snow or ice, the relationship is rather more abrupt, and the maximum adhesion occurs at higher slip values.

The braking action may be determined in several ways:

✈ by estimation

✈ by measuring the distance or time required for the vehicle to come to a full stop.

✈ by braking the vehicle equipped with an accelerometer

✈ by special devices for continual measurement (Skiddometer, Surface Friction Tester, Mu-meter, Runway Friction Tester, GRIPTESTER).

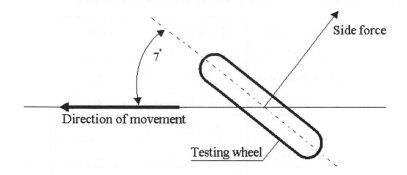

Figure 5-17   The principle of measuring friction coefficient by Mu-meter

The data obtained from the measurements carried out by particular devices cannot be used directly to quote a braking action. In order that the braking action can be reported, operational experience is also needed to interpret them.

When the braking action is determined on the basis of measurement of the distance or time required for complete stoppage of the braked vehicle, the testing vehicle is accelerated to a specified speed. The vehicle is usually occupied by two persons. At a chosen moment, the driver locks the brakes, giving 100 per cent slip. Therefore the data obtained by that method are worse than the real conditions.

The deceleration meter may only be used under certain conditions, e.g. on a runway covered with compacted snow, ice or a very thin layer of dry snow. The deceleration meter or a braked vehicle should not be used when the runway surface covering is not continuous. Reliable reporting of the braking action may be obtained from those devices that measure friction at high speed. They each function on different principles.

The Mu-meter (skid meter) evaluates a braking coefficient from the value of lateral force of the tested wheel mounted under an angle of 7° towards the movement of the vehicle moving at a speed of 66 km.h^{-1} (± 4.8 km.h^{-1}).

Saab Friction Tester (slip meter) represents an application of the anti-locking equipment used in the aeroplanes. The braking coefficient is determined on the basis of the value of peripheral force (braking force) measured at a tested wheel with the slippage adjusted to a constant 15 %. The results from the measurements at a speed of 130 km.h^{-1} approximate to real conditions of an aeroplane moving on the runway.

Table 5-5   Friction data and descriptive terms of braking action

Measured coefficient	Estimated braking action	Code
0.40 and above	Good	5
0.39 to 0.36	Medium to good	4
0.35 to 0.30	Medium	3
0.29 to 0.26	Medium to poor	2
0.25 and below	Poor	1

The results from these methods are clearly not directly comparable. Therefore it is always necessary to state by which method, device and at which speed the braking action was determined.

Table 5-5 includes the data on the measured or calculated braking coefficients and their coding. The data were determined on the basis of experience and measurements with one type of deceleration meter and one type of continual measuring device at constant speeds. The braking action must be reported for every third part of the runway separately.

Figure 5-18   Pavement transversal grooving (photo A. Kazda)

6

AIRPORT SITE SELECTION AND RUNWAY SYSTEM ORIENTATION

Tony Kazda and Bob Caves

6.1   SELECTION OF A SITE FOR THE AIRPORT

The construction of a new aerodrome or an enlargement of an existing one represents extensive investments and building works. It is therefore necessary to design the entire aerodrome project for the longest time period possible. The maximum possibilities of the airport development in the proposed locality should be considered, within the limits of the airport's critical constraints. As well as ensuring that the capacity and operational requirements are met safely, the issues concerning the airport and its surroundings should be considered, particularly the impact of the airport on the nearby population and environment. The locality selected for the airport and orientation of the runway system should facilitate a long-term development of the airport at the lowest cost in terms of money and social impacts.

Selection of a suitable site for the airport should begin with an assessment of any existing airport and its site. It is nearly always easier to modify an existing airport than to create a new one on land that has previously had a different land use designation. The assessment is made in the light of the prospective passenger market, its growth rate and any limitation of the growth resulting from a demographic shift of population. Therefore the prognosis of the growth of a number of passengers and volume of air cargo in the catchment area of the airport is one of the key elements in planning the airport development. The methodology of forecasting traffic is considered in Chapter 18.

After the proposed airport's size and layout has been approximately determined by a preliminary study, areas for possible development of the airport are assessed in several steps, the principal ones being:

→ approximate determination of the required land area

→ assessment of the factors affecting the airport location

→ preliminary selection of possible localities from maps

→ survey of individual sites

→ assessment of impacts on the environment

→ revaluation of the selection of possible sites

→ production of site layout drawings

→ estimate of costs, revenues and discounted cash flow

→ final selection and assessment of the preferred site

→ elaboration of the final report and recommendations.

The same procedure is followed in principle when assessing the development of an existing airport, though some of the steps may be omitted.

The number, location and orientation of the runways and of the taxiway system should meet the following criteria:

→ the availability (the usability factor) of the runway system should be acceptable from an economic viewpoint in comparison with the losses which would result from the unavailability of the airport

→ obstacle restrictions defined by the obstacle clearance limits should be respected

→ the final capacity of the runway system should meet the predicted demand in the typical peak hour traffic in the far future

→ the area selected for the airport should be considered not only from the viewpoint of obstacle clearance limits and environmental requirements but also from the viewpoint of the location of aerodrome facilities

→ the ground transport access to the airport and the urban development plans in the airport vicinity should be efficient and sustainable.

6.2   USABILITY FACTOR

The ease with which acceptable usability of the airport can be provided depends on meteorological conditions in the selected locality and the topography of the site. The usability factor is defined in Annex 14 as: 'The percentage of time during which the use of a runway or system of runways is not restricted because of the cross-wind component.' It can therefore be provided most economically by optimising the runway directions relative to the prevailing wind directions. The actual runway orientation is the result of a compromise between the usability and other factors. The sensitivity of an aeroplane to the cross - wind depends mainly on the aeroplane mass and the type of undercarriage. The greater the aeroplane mass, the faster it tends to fly and the less sensitive it is to the wind. Airports for large aircraft should therefore be able to meet the usability requirements without resorting to more than one direction, but the design will also be influenced by the length of runway to be provided, the associated difficulty of leveling the site and the obstacle clearance requirements. In any case, it is usually necessary for a large airport also to cater for the needs of smaller aircraft, so it will often be necessary to install short cross runways to provide adequate usability for them.

Table 6–1   Wind distribution statistics by direction and speed in per mille
Meteorological station – Bratislava M.R. Štefánik airport

Wind speed/ direction	N	NE	E	SE	S	SW	W	NW	calm	Σ
calm	-	-	-	-	-	-	-	-	123	123
1 - 2	40	82	52	31	25	23	23	65		341
3 – 5	45	45	34	37	26	16	19	111		333
6 – 10	39	3	5	15	16	4	9	90		181
11 – 15	4	0	-	0	1	0	1	15		21
16 – 20	-	-	-	-	-	-	-	1		1
> 20	-	-	-	-	-	-	-	-		-
Σ	128	130	91	83	68	43	52	282	123	1000

The number of runways in the airport and their orientations should provide at least 95 % usability factor for the types of aircraft which are intended to use the aerodrome, presuming very conservative crosswind capability of the aircraft. A basis for calculation of the usability factor are the data on percentage occurrence of the wind of the given direction and speed for a period of at least five years. The statistical data are usually published in the form of a table (see Table 6 - 1), or in the form of a compass rose.

Determination of the usability factor may be made by several methods. It is best to choose a method which is likely to underestimate the usability rather than the opposite.

For a rapid assessment, the transparency method is used (Figure 6-1). The statistical data on percentage occurrence of the wind of each combination of direction and speed are plotted in the form of a compass rose. The value of percentage occurrence is recorded in a window which corresponds to the direction and speed of the wind. The width of the overlaid transparency is equal

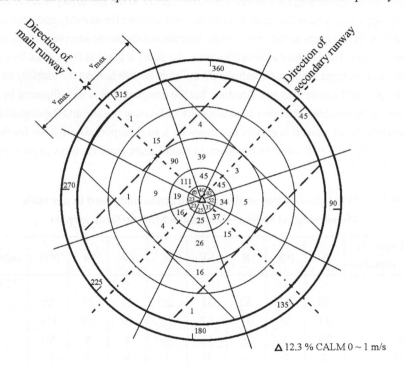

Figure 6-1   Determination of the usability factor - transparency method,
Bratislava M.R. Stefánik Airport

to twice the maximum allowable cross-wind speed. The transparency is turned in such a way that it covers the maximum of the percentage wind occurrences. The same method may be repeated for each additional runway until the combined coverage meets the required usability factor of the runway system. The whole process is then repeated for other maximum crosswind criteria.

For a more exact evaluation of the usability factor, it is convenient to use the Durst method. The basis of this method is a rectangular co-ordinate system. Directions of a compass rose are developed in the x axis and speeds of the wind in the y axis. In the created fields the percentage wind occurrence and speed are recorded. The following function is plotted into the grid:

$$f = v \cdot \frac{1}{\sin \alpha}$$

where:

v   is the maximum value of the cross-wind

α   is the angle which is formed by the runway axis with particular directions of the compass rose.

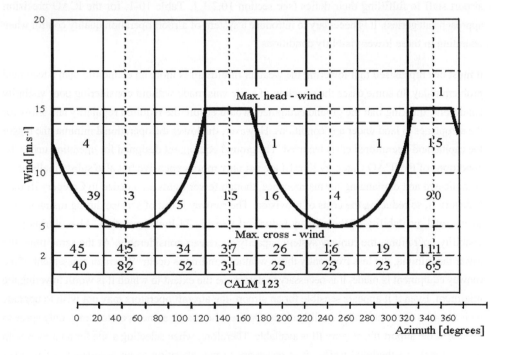

Figure 6-2   *Determination of the usability factor – Durst method,*
Bratislava M.R. Stefánik Airport

In the direction of the runway axis, the function increases to infinity and the usability factor is limited by the maximum value of the head wind. The lowest point of the function is actually a value of the maximum cross-wind (α = 90°). When the numerical values under the function are summed, the usability factor of the runway, or of the runway system is given.

6.3   USABILITY FACTOR WITH THE RUNWAY VISIBILITY RANGE

Large airports are expected to continue to operate in poor weather but it is important to consider the occurrence of low visibility and low ceilings of the cloud base. The occurrence of Category II or III visibility represents a significant qualitative change as a consequence of considerably lowered abilities of the pilots, air traffic controllers and other persons involved to maintain the procedures and a expeditious flow of air traffic with a visual reference. A change from I to II and any lower category is more than just the provision of the respective technical equipment and appropriate limitations. It requires also a change of the system of airport control and of the attitude of all the airport staff to fulfilling their duties (see section 10.2.1.1, Table 10-1, for the ICAO precision approach categories). It is necessary to introduce a system of airport operation quality control when operating in these lower visibility conditions.

It must be emphasised that lowering the weather minimum is more an economic than a technical problem today. In some cases the selection of a site was made without considering poor visibility conditions, thinking that the ground equipment would ensure the required regularity and allow for the aeroplanes to land under any conditions. However, the lower the operational minima, the bigger the capital and operational costs incurred. The ground equipment designed for operation under the conditions of the ICAO Category II or III are not only more expensive but also substantially more complicated and demanding for maintenance, though fewer lights are needed in Category III then Category II, so reducing the costs of provision. The runway is out of service during maintenance, so the total availability of the runway reduces if there is 24 hour operation. Also, during poor visibility operations, the runway system capacity decreases considerably. At the same time, the lower the minima, the less frequently those weather conditions occur. When an assessment of the runway equipment is made, it is necessary to determine the extent to which it is worth lowering the minimum. Even if it appears sensible for an airport, the aircraft operators may not wish to upgrade their avionics fit or their pilot qualifications. On the other hand, some operators may only agree to operate at the airport if Category III is available. Therefore, when selecting a site for an airport with international or scheduled traffic, it is necessary to pay attention to an assessment of weather conditions, including occurrence of reduced visibility and low cloud base, and the associated costs and benefits. Fortunately, new GPS-based navigation may, by 2020, allow these low visibility operations to be performed with less investment by individual airports.

For determination of the usability factor including the effect of low visibility, the Antonín Kazda method may be employed./1 The method evaluates the usability factor in two steps:

1. Evaluation of the usability factor including the effect of crosswind if the visibility or the runway visual range is greater or equal to 800 m.

2. Evaluation of the usability factor considering the RVR limits without taking the wind into account if the visibility or the runway visual range drops under 800 m.

This two step method may be employed because of a mutual dependency of the occurrence of low visibility and low velocity winds, i.e. 'when there is a fog, no strong wind blows', has been effectively proven with a high probability.

The evaluation of the usability factor with the wind effect may be performed by any of the known methods, for example the Durst method described above.

For evaluation of the usability factor considering the runway visual range limits, the method of least squares, the approximation of accumulated occurrences of visibility by the means of a polynomial, is employed. The method allows the usability factor to be evaluated with respect to the visibility limits in an arbitrary season and a time interval of a day, and may describe the time dependency (daily and yearly) of occurrences of low visibility.

The resultant usability of the aerodromes is the sum of the loss of usability due to the wind effect and the loss of usability due to the runway visual range limits.

6.4   OBSTACLES

In the complex orographical conditions of mountainous countries, the appropriate runway orientation to minimise the crosswind is often coincident with the orientation designed with regard to obstacles. This is likely because the direction of the prevailing winds often coincides with the valley centre line. Even so, the obstacle limitation surfaces may well be penetrated by obstacles. Development of many airports, particularly extending the runway, is often difficult because of obstacles. The defined obstacle-free airspace is an inseparable part of each aerodrome and should permit the intended aeroplane take-off, approach as well as landing and operation at the aerodrome circuit to be conducted safely.

---

1 KAZDA A.: Evaluation of the Usability Factor of Aerodromes. Candidate dissertation thesis, University of Transport and Communications Žilina 1988

The airspace round the airport is defined by a system of obstacle limitation surfaces. The characteristics of obstacle limitation surfaces are specified on the basis of types of airports (transport, sports, heliports, etc.) and are related to the intended use of the runway in terms of take-off, landing and the type of approach (non-instrument approach, instrument approach, precision approach).

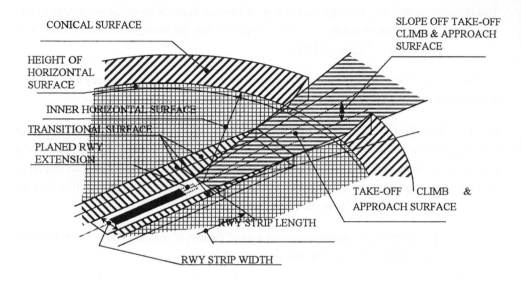

Figure 6-3   *Obstacle limitation surfaces*

In cases where operations are conducted to or from both directions, then the function of certain obstacle limitation surfaces or parts of them may be nullified because of more stringent requirements of other lower obstacle limitation surfaces. Dimensions and slopes of obstacle limitation surfaces were derived from the performance characteristics of the various types of aeroplanes and from the statistical probability of deviations of the aeroplanes' trajectories relative to the intended flight path. Characteristics of obstacle limitation surfaces correspond to the requirements of the regulations for flight operations, air traffic control and aircraft airworthiness. Even in the most extreme foreseen position of the aeroplane, the prescribed minimum horizontal and vertical distance of the aeroplane from obstacles should be maintained. The system of obstacle limitation surfaces as presented in Annex 14 also serves to advise a local authority of the height limits for new objects that might restrict or endanger the operations of aeroplanes in the future. If there is a potential conflict, the relevant aviation authority should be consulted.

For non-instrument runways, for instrument approach runways and for precision approach runways of Category I, the following system of obstacle limitation surfaces has been specified:

+ conical surface

+ inner horizontal surface

+ approach surfaces

+ transitional surfaces

+ take-off climb surfaces.

The most significant differences in characteristics between these classes of runway are related to the dimensions and slopes of the take-off climb and approach surfaces. The necessary data may be found in Annex 14. In Table 6-2 the dimensions and slopes for precision approach Category II or III are specified.

The system of obstacle limitation surfaces for precision approach runway Category II and III is supplemented by:

+ inner approach surface

+ inner transitional surfaces

+ balked landing surface.

These come into play when operations are conducted in the relevant low visibility conditions. The determinant for assessing the consequences of a certain obstacle penetrating a surface is the type of operation on the runway in question, the position of the obstacle (which surface is obstructed by the obstacle, the distance from the runway threshold/end and how much it obstructs it) and its character (hill, aerial transmission line poles, road profile, etc). Depending on these factors, the Aviation Authority may still allow operations on the basis of a study, often using the ICAO Collision Risk Model.

Obstacles in the take-off climb and approach surfaces and in the transitional surfaces are assessed most stringently. The construction of new objects or extensions of existing objects shall not be permitted if they would penetrate those surfaces. It is, however, not always possible to eliminate the occurrence of obstacles. Then it is necessary to determine special procedures such as offset approach or take-off trajectory , or install special equipment for the runway, or limit the runway operation with higher operating limits as well as marking and lighting the obstacles.

If the obstacles only penetrate the conical surface or the inner horizontal surface, less stringent criteria are used. Even in that case, for an assessment of a specific obstacle, the location and character of the obstacle are important.

Table 6-2 Dimensions and slopes of obstacle limitation surfaces for precision Category II or III RWY

SURFACE	DIMENSIONS
CONICAL	
✈ Slope	5 %
✈ Height	100 m
INNER HORIZONTAL	
✈ Slope	45 m
✈ Height	4 000 m
INNER APPROACH	
✈ Width	120 m
✈ Distance from THR	60 m
✈ Length	900 m
✈ Slope	2 %
APPROACH	
✈ Length of inner edge	300 m
✈ Distance from THR	60 m
✈ Divergence	15 %
First section	
✈ Length	3 000 m
✈ Slope	2 %
Second section	
✈ Length	3 600 m
✈ Slope	2.5 %
Horizontal section	
✈ Length	8 400 m
✈ Total length	15 000 m
TRANSITIONAL	
✈ Slope	14.3 %
INNER TRANSITIONAL	
✈ Slope	33.3. %
BALKED LANDING SURFACE	
✈ Length of inner edge	120 m
✈ Distance from threshold	1800 m
✈ Divergence	10 %
✈ Slope	3.33 %
TAKE-OFF CLIMB	
✈ Length of inner edge	180 m
✈ Distance from RWY end	60 m
✈ Divergence	12.5 %
✈ Final width	1 200 (1 800) m/[1]
✈ Total length	15 000 m
✈ Slope	2 %

1/ 1 800 m when the intended track includes changes of heading greater then 15° for operation conducted in IMC, VMC by night.

It is necessary to prove whether the obstacle is shielded by an existing immovable object, or whether it is determined after aeronautical study that the object would not adversely affect the safety or the regularity of operations of aeroplanes.

Some objects may be considered particularly dangerous and should be removed or at least marked even if they do not obstruct any obstacle limitation surfaces. They are, in particular, isolated thin objects such as chimneys, poles and posts, or aerial high and extra-high tension transmission lines in the approach and take-off climb surfaces. The Aviation Authority may order removal (e.g. trees) or marking (e.g. aerial high tension transmission line) of any object that might after aeronautical study endanger aeroplanes on the movement area or in the air. The object should generally be removed by its owner, not by the aerodrome operator.

6.5   OTHER FACTORS

Though the system of take-off and landing runways and taxiways represents the greatest demands for space, the options are often very limited. Therefore, in planning an airport, the take-off and landing runways and taxiways are usually designed first. The capacity of the runway systems is a limiting factor of further development of the largest airports. In many cases, the approval process of the construction of a new runway or extending one may take tens of years, and further expansion of the airport is often impossible due to public opposition.

The airspace capacity of the control area and of the runway system of the airport may be assessed with computer simulations, analytical models or handbook methods. SIMMOD is a simulation package for assessing the capacity of the airspace and the movement areas of the airport, and FAA Airfield Capacity Model is a handbook method for assessing the runway and the maneuvering areas. According to the FAA, it is possible to define the runway capacity as the maximum number of movements within an hour once the tolerable delay has been defined.

The magnitude of an acceptable delay differs at individual aerodromes. From the maximum number of movements in an hour, it is possible to derive the daily and yearly capacity of the runway system. Figure 6-4 illustrates how the capacity varies with the runway configuration.

Figure 6-5 illustrates how it is possible to grow the capacity of the airport in terms of the number of possible movements in a year.

On one runway with a optimum layout of taxiways and with a sufficient capacity of the apron, it is possible to reach up to 250 000 movements in a year, as at London Gatwick in 1998, even using Instrument Flight Rules (IFR).

Number	Runway use configuration	Hourly capacity ops/h		Annual service volume
		VFR	IFR	
1		51 - 98	50 - 59	195 000 - 240 000
2	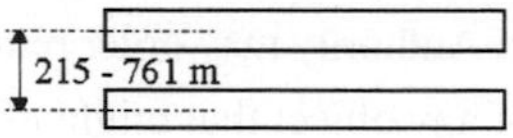 215 - 761 m	94 - 197	56 - 60	260 000 -355 000
3	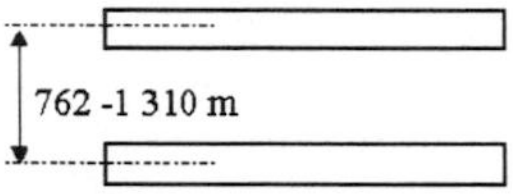 762 -1 310 m	103 - 197	62 - 75	275 000 - 365 000
4	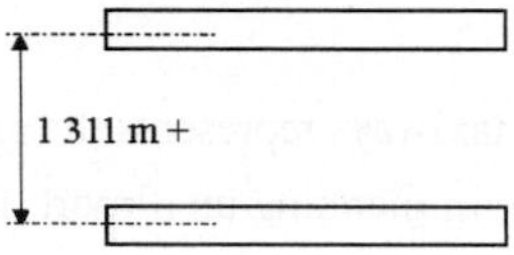 1 311 m +	103 - 197	99 - 119	305 000 - 370 000
5	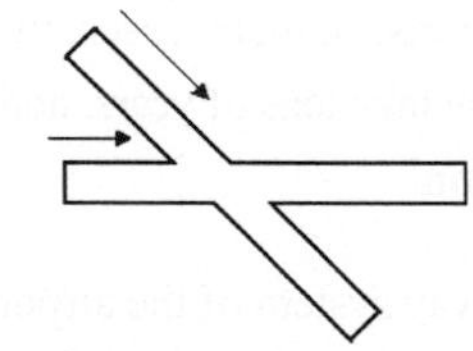	72 -98	56 - 60	200 000 - 265 000
6	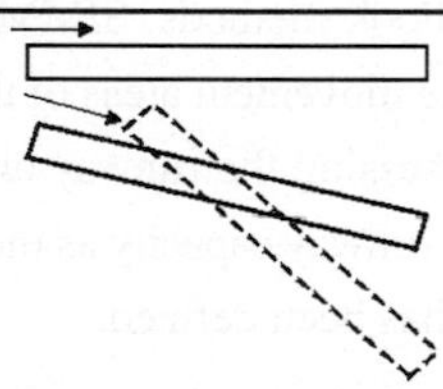	73 - 150	56 - 60	220 000 - 270 000
7	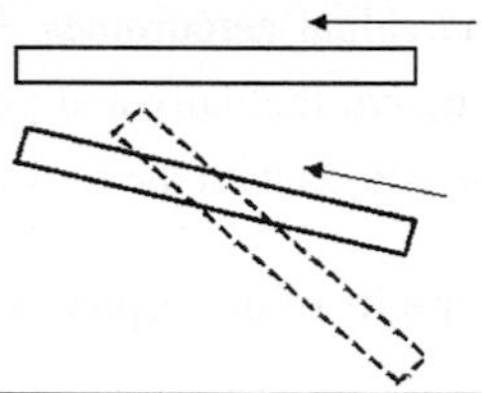	73 -132	56 - 60	215 000 - 265 000

Figure 6-4   *Hourly capacity and annual service volume*
Source: ICAO Airport Planning Manual

Of these, 243 000 were Air Transport Movements. However, a decision to construct another runway depends not only on the annual runway capacity but also on the magnitude of the peak hour traffic and the type of operation ('hub and spoke' or 'point to point'), as well as on an assessment of other factors which, if they occurred, would bring about a complete shut down of an airport with one runway. These might be an accident, an unlawful act of intervention in an airport or an aeroplane, winter maintenance of the runway, etc. The design of systems of taxiways is dealt with in Chapter 4.

The development of airports and further growth of air transport is often limited by a so called environmental capacity, or carrying capacity, of airports. The term environmental capacity means a limitation of air traffic or of a further development of the airport in relation to the effects of noise, air pollution, underground waters and soil, visual pollution, etc.

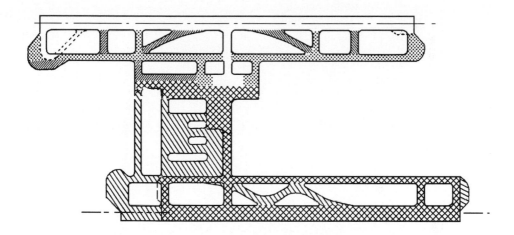

Figure 6-5 Typical phased development of airport

Source: ICAO Airport Planning Manual

Level at which to build

☐ Up to 20 000 to 30 000 operations

▦ 30 000 to 60 000 operations

▨ 50 000 to 99 000 operations

▩ 75 000 to 150 000 operations

▨ 150 000 to 250 000 operations

At many airports there is a prohibition or a limitation against operations at night. These night curfews as in Tokio's Narita, cause further intensification of peak hour traffic or a limitation of the number of movements with regards to the noise load from particular types of aircraft as in Munich,

Düsseldorf and Frankfurt. There are also limitations of the number of movements with regard to the level of air pollution as in Zürich and Stockholm's-Arlanda, and other administrative limitations. Chapter 17 of this book deals with the problems of environment protection and ecological limitations.

7

APRONS

Tony Kazda and Bob Caves

7.1 APRON REQUIREMENTS

Aprons are designed for parking aeroplanes and turning them around between flights. They should permit the on and off loading of passengers, and technical servicing of aeroplanes including refuelling and the on-and off loading of baggage and cargo.

The requirements for the construction of aprons are similar to those of the other reinforced surfaces. As was mentioned in Section 5.3, the aprons are the most loaded of all the movement area pavements. The aeroplane has its maximum mass on the apron just prior to departure. The surface is subject to concentrated point loads from the wheels of standing or slowly moving aeroplanes. In addition, it is stressed by dynamic vibrations after starting up the engines. Therefore, when dimensioning the apron pavement thickness, a safety factor 1.1 is used.

In the past, concrete was used almost exclusively for apron pavements. The original concrete aprons are often renovated by covering them with layers of asphalt. On aerodromes intended for use by small or medium–sized aeroplanes, asphalt is commonly used for aprons. As already mentioned in Section 5.2.3.5, the use of block paving for aprons has also expanded.

In order that aeroplanes can be park and taxi out under their own power, the slopes should be minimised but still be sufficient to allow the surface water to drain adequately. In the aircraft parking area, the maximum slope of the apron should not exceed 1 %.

The apron is a bridging point between the runway system and the terminal building. The location of the apron and its aircraft stands should allow convenient access. The following are some of the basic requirements that should be objectives when designing the apron:

+ location of the apron to minimise the length and complexity of taxiing between the runway and the stands

+ the apron should permit mutually independent movements of the aeroplanes on to and off stands with minimum delay

+ on the apron it must be possible to locate a sufficient number of stands to cope with the maximum number of aircraft expected during the peak hour

+ the apron should be adequate to allow quick loading and unloading of passengers and cargo

+ the aprons should be designed so that there is sufficient space for the turnround activities to be performed independently of activities on an adjacent stand.

+ the apron area should be adequate to provide sufficient space for parking and manoeuvring of the handling equipment , and also for the technical personnel

+ there should be a safe and effective system of airside roads for technical equipment to access the stands.

+ the negative impacts on the workers' environment, particularly safety, noise and exhaust gases should be minimised, the emphasis being the health and safety of the staff

+ the possibility of further extension of the runway system, aprons and buildings should be considered.

Each of these requirements has an effect upon the final design of the apron.

7.2   APRON SIZING

The above objectives can only be achieved if the apron size and the positioning of the stands is adequate to permit expeditious handling of the aerodrome operation during predicted peak hour traffic levels.

The appropriate apron size depends on the types of aircraft which are intended to use the apron. Each aircraft type needs sufficient stands in the correct position, and possibly a specific manner of stand operation, either nose-in or inclined for self-manoeuvring. There may also be specific requirements for technical servicing (e.g. refuelling), for technical equipment, and for guidance to the parking position when nose-in operation is used. The accuracy of the aircraft guidance on the apron may have an effect upon the size of the apron.

The total apron area depends also on the type of the operation prevailing on the aerodrome and the occupancy time of stands. For some airline operators, the aerodrome may serve as a base, for others, as a departure/arrival aerodrome or only as a transit stop. The based aircraft normally have a greater occupancy time between arrival and departure, so much so that it may be better to consider towing them off to remote stands.

In order that all the factors that influence the apron size may be considered, particularly with large aerodromes, it is advisable to simulate the operations on the apron. Similarly, the operational control of the apron is computerised.

7.3   APRON LOCATION

Theoretically, the most efficient location for the apron is at a distance of 1/3 length of the runway from the main runway threshold, (Figure 7-1), based on a common split of use of the two runway thresholds.

When determining the apron location and orientation, the following objectives should be taken into account:

✈ minimum length of taxiing of the aeroplanes

✈ the shortest distance possible from the walkway in front of the terminal building to and from the aeroplane

✈ minimum impact of the engine exhaust on the terminal building

✈ possible need to extend the apron as well as the terminal building.

In the majority of cases, the apron is constructed directly in front of the terminal building, allowing the stands to be served by airbridges if they are deemed to be necessary. If bridges are not provided, some other safe way of getting the passengers to and from the aircraft will be needed, using well-policed walkways or buses. Either the airside service road runs under the airbridges, or it has to run between the back of the stands and the taxiways.

In some other cases, the whole apron, or a section of it, is constructed at a considerable distance from the terminal building. Then it is necessary to provide some form of transport of passengers between the terminal building and aeroplanes. This increases the operating costs. There are many reasons why an apron might be located remotely.

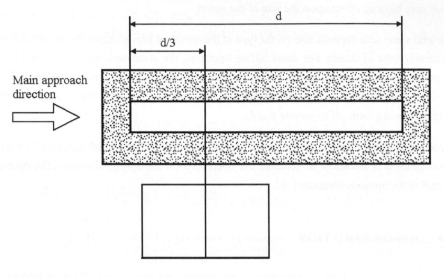

Figure 7-1   Apron location

It may be to try to create a comfortable ambience in the terminal building and to reduce walking distances as with the mobile lounges at Montreal, a need to earmark a section of the apron for longer-time parking of aeroplanes as at Washington D.C., to generate the space to handle large aircraft as at Los Angeles, or to provide a cost-effective way of handling the additional peak hour traffic as at Munich or Rio Galleao. In the majority of cases, it has been the practice for the airport operator to allocate the contact stands (those directly adjacent to the terminal building) for high-capacity aircraft and scheduled airlines, particularly if they are part of a hubbing operation.

7.4   APRON CONCEPTS

The geometric and manoeuvring characteristics of aeroplanes make it practically impossible in most cases to locate all the stands required for peak traffic directly adjacent to the central processing part of the terminal building. It is therefore necessary to generate other solutions. Several basic concepts that have developed over time may be identified, depending on the total size of the airport. Each concept has its advantages and disadvantages, so the solution is often a combination of the basic concepts discussed below.

7.4.1 Linear Concept

This is the simplest layout. Individual stands are located along the terminal building (Figure 7-2). as at Roissy-Charles de Gaulle-Aérogare 2 and 3, or around the terminal building as at Birmingham's Eurohub.

A modification of that concept may be found in large airports where the stands are placed along several parallel passenger loading piers that are connected with one another and also with the central terminal by a transport system, as at Atlanta Hartsfield, and planned for Heathrow-Terminal 5.

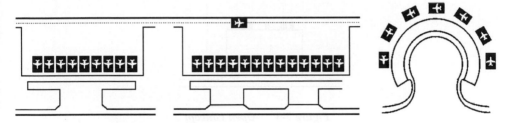

Figure 7-2   Linear concept
Source: Airport Planning Manual, Part 1 Master Planning

An advantage of the linear apron concept is the simple access from the terminal building to the aeroplanes, a simple installation of the passenger loading bridges and sufficient space for technical handling equipment and staff at the level of the apron. A disadvantage in larger airports may be the large distance between the extreme stands and the terminal building, and sometimes an even larger distance to another stand for transfers between airlines. The latter problem may be solved by people movers. There is now pressure on airports to rearrange the stand allocations to ensure short transfer distances between flights within an airline grouping, so giving the carrier's hub a competitive advantage.

7.4.2 Open Concept

In this concept, the stands are located on one or more rows in front of the building (Figure 7-3).

One of the rows may be close-in, but most will be a long way from the terminal. The transport of passengers to the distant stands is provided by buses or mobile lounges. Athens West terminal and Milan Linate are examples where almost all the aircraft stands are on open aprons. The concept allows many aircraft to be served from a very short terminal frontage.

A main disadvantage is the need to provide transport to distant stands, requiring a large workforce and fleet of buses. The length and lack of reliability of these bus trips makes the concept unsuitable for operations with transfer passengers. Another is the large number of additional movements on the apron, increasing the possibility of accidents with aircraft and other ground vehicles.

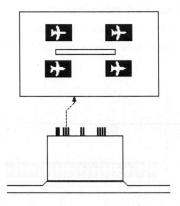

Figure 7-3   Open concept
Source: Airport Planning Manual, Part 1 Master Planning

7.4.3   Pier Concept

In many large airports, the introduction or extension of piers were practically the only possibility for providing a greater number of contact stands and to increase the capacity of the airport. The shape of the passenger loading piers varies and depends on the space available at the airport (Figure 7-4). London Heathrow airport, shown in Figure 7-5, is a classic example of such a solution. Piers have the advantage of keeping the footprint of the terminal and apron complex quite compact and, most importantly, keeping all the gates under one roof. This allows direct contact with the central processing area and a relatively simple navigating task for transferring passengers.

Piers generally involve the aircraft having to taxi into cul-de-sacs to get to the stands. The limited space leaves little extra room for aircraft handling. There is usually only a small parking bay for equipment and service personnel under the piers. In the confined space, the effects of noise and exhaust fumes create poor working conditions for the staff. Any increase in aircraft size tends to result in congestion on the cul-de-sac taxilanes.

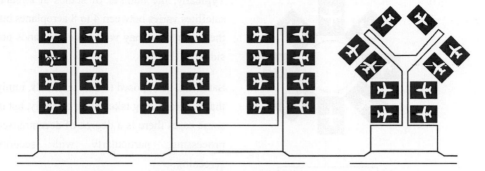

Figure 7-4   Pier concept

Source: Airport Planning Manual, Part 1 Master Planning

Figure 7-5   Pier system at London Heathrow Airport (photo A. Kazda)

7.4.4   Satellite Concept

In this concept, each of the remote passenger loading satellites are connected with the terminal building by underground tunnels or by overhead corridors, as in Figure 7-6. The satellites may be any shape from linear as at Atlanta Hartsfield to circular as with Charles de Gaulle Terminal One.

Typically, the number of stands at circular satellites varies between 4 to 8 aeroplanes but the linear ones may well have 20 stands per side.

Satellites, as opposed to unit terminals, imply that the processing takes place centrally, but in some cases there is a degree of decentralised processing, particularly with security screening.

The satellite concept avoids cul-de-sacs and their disadvantages. However, it requires a larger total apron space. Since the distance from the terminal building to the aeroplane is considerably extended, it is necessary to transport passengers between the terminal

Figure 7-6   Satelite concept
Source: Airport Planning Manual, Part 1

building and the satellites, usually with a high frequency automated people-mover.

7.5   STAND TYPES

Individual stands may be designed as 'taxi out', sometimes referred to as 'self-manoeuvring'. The aeroplane taxis in and out under its own power.

The alternative is a 'nose-in' stand. The aeroplane taxis in under its own power, but to leave the gate it must be pushed back into a specific position usually through 90 degrees onto the adjacent taxilane. In most cases, the engines may not be started until the push-back is complete.

Taxi out aircraft stands are mostly used at small airports with a small number of movements. Advantages of the taxi out aircraft stands are their low operating costs, though they do require a marshaller, together with their flexibility. In the majority of instances, daily inspection of the marking of the apron is sufficient. The type of operation makes it almost impossible to use a passenger loading bridge. A disadvantage is the greater space requirement. Approximately 10 stands of the nose-in type may be positioned in the same space as 8 taxi out aircraft stands, thus increasing the apron utilisation. The size of the taxi out aircraft stand depends on the turning radius, wing span and the aeroplane fuselage length of the aircraft. When the aeroplane turns, a so called safety distance between the wing tip and an obstacle should be ensured, as shown in Figure 7-8. The stands may overlap one another but, if they do, simultaneous movements of the aeroplanes on adjacent stands are not possible.

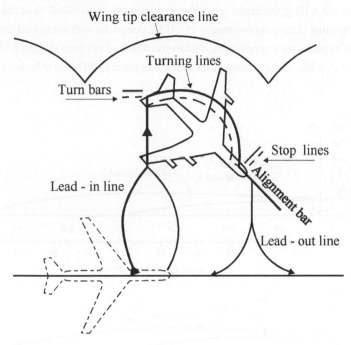

Figure 7-7   Aircraft stand markings

When the aeroplane is turning, particularly if it has a multi-wheel main undercarriage, there is a considerable shear stress on the pavement.

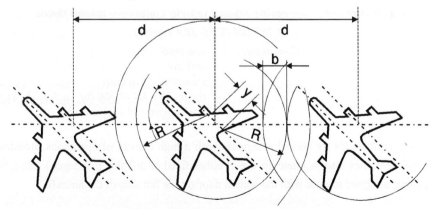

Figure 7-8   Individual stands overlap

Pavements with an asphalt surface are often damaged in this way. Another disadvantage of the power-out stands is in negative impacts on the terminal building, other objects, or the traffic on the apron produced by the engine's exhaust plume at brake-away thrust and during turning. In order to start the aeroplane rolling, the brake–away thrust represents up to 50-60 % of the maximum continuous thrust, while turning approximately 25-30 %, compared with normal taxiing when, for aircraft with high bypass ratio engines, there is often too much power even at ground idle settings. Figure 7-9 gives velocities of the exhaust gases of the aeroplane B-767-200 at brake–away on the stand.

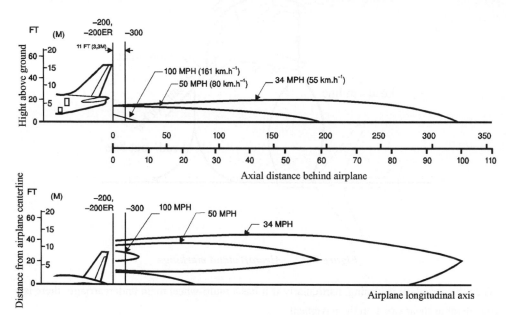

Figure 7-9   *Jet engine exhaust velocity contours – takeoff thrust*
B 767-200 (CF6-80A engine)

Conditions: - sea level
- still air
- standstill aeroplane
- thrust 38.6 kN (8 500 lb), each of the engines
- both engines in operation

Nose-in aircraft stands are normal at airports with a high density of operations. Besides being space-efficient, they also facilitate the use of either fixed or drivable passenger loading bridges, these being designed to mate with the second door on the left side of the aircraft.

They cost more to equip and operate than the self-manoeuvring type. As well as the air bridge, if it is provided, there are the tractors and parking guidance systems to be provided. There may well also be hydrant fuel systems and fixed power installations, which are more appropriate than for the more flexible open-apron power-out operations.

The details related to technical servicing of aeroplanes on both types of the stands are given in Chapter 8.

On the stands, the aeroplanes may be positioned nose inwards, at an angle nose inwards, nose outwards, at an angle nose outwards, and parallel to the terminal building, as shown in Figure 7-10. Each solution has its advantages and disadvantages.

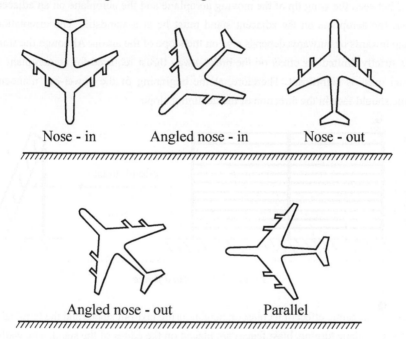

Nose - in   Angled nose - in   Nose - out

Angled nose - out   Parallel

Figure 7-10   Positioning of aeroplanes on the stands

On the nose in–push back aircraft stands, the aeroplanes are located either directly nose inwards, or at an angle between 30° to 60°. The angle depends on the size, shape and layout of the apron. The fundamental requirement is for an aircraft on stand not to block the apron taxiway behind it, and for the tow tractor not to block the service road in front of the aeroplane when initiating push-back.

When power-out operations are used, the aeroplanes may be theoretically parked in any of the above ways. The 'nose inwards' or the 'nose outwards' placement, i.e. right-angled to the terminal building edge are normally not used because the stands cannot overlap one another, and thus the aeroplanes occupy the maximum area. Placing the aeroplane parallel to the terminal building is convenient particularly due to the fact that the passenger loading bridges may be installed to both the front and the rear door of the aeroplane. A disadvantage is that after starting up the engines, it is practically impossible for the aeroplane on the stand behind to be serviced . Therefore this parking arrangement is also used only rarely, and only at airports with a sufficiently large apron, a low frequency of flights or with the circular satellite stand concept. Aircraft are most often parked at an angle of 30° up to 45° inwards or outwards from the terminal building frontage. In spite of the fact that individual stands overlap one another, the prescribed safety distance should be provided between the wing tip of the moving aeroplane and the aeroplane on an adjacent stand. However, the aeroplane on the adjacent stand must be at a standstill. The orientation of the aeroplane inwards or outwards depends also on the slope of the apron. Although the slope of the apron is strictly limited, its effect on the brake–away thrust required for an aeroplane with the maximum mass is significant. Therefore, at the beginning of the power-out manoeuvre, the aeroplane should face in the direction of the declining slope.

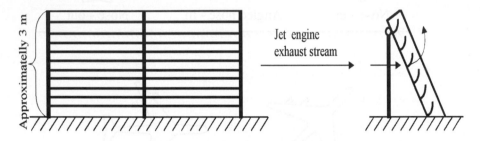

Figure 7-11   Blast fence

To relieve the dynamic effect of jet blast on neighbouring taxi-out stands and the terminal building during the aeroplane turning, blast fences are placed on the edges of the stand. The walls usually consist of sets of steel vanes which turn the jet exhaust upwards, as shown in Figure 7-11. The height of the walls on the apron is limited and in most cases does not exceed 3 m. Because of this they are effective for under wing mounted engines only.

7.6   APRON CAPACITY

At a properly dimensioned aerodrome, the capacity of all parts of its system are in approximate equilibrium. There are then no critical bottlenecks in the system. Some portions of the aerodrome can bear a short-term overload, others cannot. The apron should be dimensioned in such a way that peak hour delay is an acceptable minimum, say no more than 2 % of flights. To cope with these small number of movements, and to ensure that delays do not back up onto the taxiways or into the air, the capacity of the apron may be increased by establishing parking ramps away from the terminal building. Such stands are usually not equipped with any technology. In the majority of cases, they are used for - charter flights and for aircraft waiting for on-line repairs. Typically, the number of parking ramps might amount to 10 % or 20 % of the number of stands on the apron, depending on the types of operation at the airport.

The number of stands on the apron depends not only on the number of movements of the aeroplanes, but also on their distribution during the day, the length of stay on stand, and on the aircrafts' dimensions and seat capacity. Theoretically, the required number of stands may be expressed by a formula:

$$N = k \cdot \frac{t \cdot n}{60}$$

where:

N   number of stands

k   coefficient of variability of use of the stand. It depends on the total number of stands and their types, and on the structure of flight timetables. It ranges between 1.3 to 2.0

t   time of aircraft turn around per aircraft movement on the runway i.e. half of turn around time in minutes

n   number of movements in peak hour.

A determination of the required number of stands by simulation is more appropriate and precise, but requires the construction of an accurate representative schedule.

7.7   ISOLATED AIRCRAFT PARKING POSITION

An isolated aircraft parking position should be established on each aerodrome for parking any aeroplane which is known or believed to be the subject of unlawful interference. The isolated parking position should be located at the minimum distance of 100 m from other parking positions, buildings or public areas. The position should not be located over underground utilities (fuel, gas, electrical, light-current cables).

8

AIRCRAFT GROUND HANDLING

Tony Kazda and Bob Caves

8.1   AIRCRAFT HANDLING METHODS

Airline companies are the most important customers of any airport. Airlines can minimise their extra investment for growth by increasing aircraft daily utilisation. A well-known phrase 'the airplane earns only when flying' holds true. Therefore, the basic requirements all airlines place on the ground handling are:

✈   to ensure safety of the aircraft, avoiding damage to it

✈   to reduce ground time

✈   to ensure high reliability of handling activities, avoiding delays.

The airport administration authorities are also interested in reducing the ground time. In order to cope with an expected growth of air transport, most large European international airports will have to invest in further development of their infrastructure. It would, therefore, be more advantageous for them to reduce the ground time and thus increase productivity without significant investments. A possible solution could be an introduction of modern technologies for aircraft ground handling. But there are many regional airports with one or only a few aircraft movements a day. When the requirements for technical handling at such airports are compared with those having busy air traffic, a considerable difference can be seen.

The process of ground handling consists of a series of highly specialised activities. In order to carry them out, both highly skilled staff and sophisticated technical equipment are needed.

Three basic approaches to ground handling can be identified:

✈ the aircraft's own technical equipment is used as much as possible

✈ mobile technical equipment of the airport is used - a 'classical approach'

✈ airport fixed distribution networks with a minimum of mobile facilities, often called a 'vehicle free apron'.

Each of the above approaches has its advantages and disadvantages. Individual approaches can also be combined. The choice of a particular approach to ground handling is basically influenced by the following:

✈ type of flight, short and long-distance

✈ availability of capacity at the airport

✈ the intensity of utilisation of a particular stand

✈ aircraft size

✈ extent of ground handling required, depending on whether it is a line or end station

✈ environmental concern of the population.

For short flights which are served with modern jet or turboprop aircraft it may well be required that the total time allotted to the ground handling during a turnaround should not exceed 10 to 15 minutes. In such cases the aircraft's technical equipment such as airstairs and the Auxiliary Power Unit (APU) is used for the ground handling as much as possible. The number of ground handling activities is reduced to minimum. Most turboprop aircraft serving short distance flights cannot use stands equipped with common passenger bridges owing to the distance of the left power plant from the bridge construction and a low door-sill height.

The BAe 146 being serviced during a line turnaround can be taken as an example of a servicing arrangement (Table 8-1).

At most medium-sized airports aircraft are serviced either by means of mobile equipment or through a combination of mobile systems and fixed ground systems. When compared with a complex installation of the fixed ground systems, the former solution requires less investment. This way of servicing an aircraft is preferred also at those airports where competition among more ground handling companies exists. In such a case only a few companies are ready to invest in ground support systems which will have to be shared with competitors, e.g. fuel systems.

Table 8-1   An example of the BAe-146 servicing at a turnaround

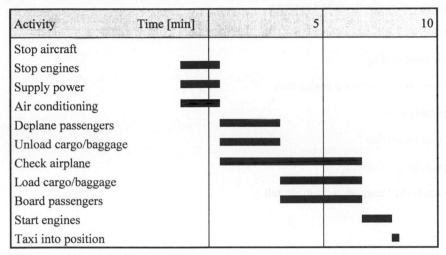

The apron at a large modern airport is not only the concrete or tarmac surface itself. In order to reduce surface traffic across the apron, to improve the apron utilisation and to reduce the ground time, the technology needed for servicing an aircraft is either buried in the surface or located under passenger bridges. The fixed ground systems considerably improve the apron safety. The rate of apron incidents is shown in Figure 8-1. The survey was conducted by ACI in November 1998 with 313 participating airports. The survey reported 671 incidents during handling of 2 133 398 aircraft movements, giving a rate of 0.315 incidents/accidents per 1 000 movements or one incident per 3180 aircraft movements. The pie graph illustrates the percentage distribution of types of incidents averaged over the period 1994-98. It shows the types of accidents/incidents which occur most frequently. Apron incidents involving aircraft are most serious. These are often caused by passenger handling or aircraft servicing equipment. Damage either to aircraft by moving equipment or to moving equipment by other equipment can be avoided by fixed services.

An apron equipped with fixed ground systems has a lower impact on the environment due to lower noise and exhaust emissions. Equally important is reduced traffic in the manoeuvring area. It also leads to saving on personnel which is significant especially in countries where the cost of labour is high. The equipment and networks buried in the apron require higher initial investment costs but, in general, reduce operational costs. It is, therefore, more suitable especially for airports with high daily utilisation of stands, but before the decision concerning a particular type of ground handling is taken, it is necessary to carry out a detailed analysis. With fixed distribution networks buried either in the apron or fixed on passenger bridges, it is possible to provide:

➤ aircraft refuelling

➤ electrical power to cockpit systems

➤ telephone and data connection

➤ air conditioning

➤ compressed air for the engine start

➤ potable water

➤ lavatory service

➤ push back of aircraft

➤ transport of baggage to/from aircraft.

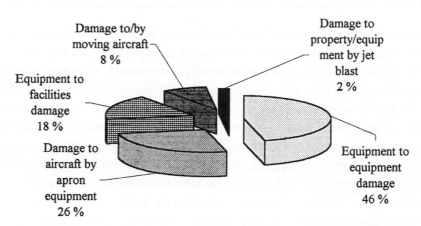

Figure 8-1   Apron incidents/accidents – causes of damage (1994-98)
Source: ACI Survey on Apron Incidents/Accidents (November 1998)

Fuel is supplied through pipes buried in the apron, while the systems on the passenger bridges provide air-conditioning, power and compressed air supplies. In the future, especially at airports which accommodate wide-bodied aircraft, a further expansion of ground support systems is to be expected. An example of a fully 'automated airport' where all systems are fixed at some stands is

Stockholm-Arlanda.

Table 8-2 presents individual technical servicing activities of the B 747 during a turnaround within a time limit of 60 minutes.

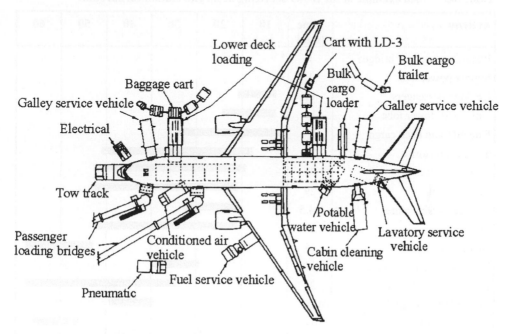

Figure 8-2   *The B-777 being serviced during a turnaround with the help of ground support systems and mobile equipment*

Source: Boeing 777 Airplane Characteristics for Airport Planning

Next, some other activities arising from aircraft servicing will be described, particularly those which implement technologies different from those used at small or medium sized airports. Aircraft refuelling will be described in Chapter 9.

8.2   AIRCRAFT GROUND HANDLING ACTIVITIES

8.2.1   Deplaning and Boarding

Deplaning and boarding can be provided by means of:

✈   stairs carried by the aircraft

→ mobile stairs

→ mobile lounges

→ passenger bridges.

Table 8-2   An example of the B-747 servicing at an end station turnaround

Activity	Time [min]	10	20	30	40	50	60
Position passenger bridges	1						
Supply power	1						
Deplane passengers	11						
Unload aft lower lobe	14						
Unload main deck cargo	25						
Service lavatories	30						
Service galleys	30						
Service cabin	29						
Service potable water	14.5						
Fuel aircraft	28						
Board passengers	18						
Unload FWD lower lobe	10						
Load main deck cargo	28						
Load FWD lower lobe	10						
Load aft lower lobe	14						
Start engines	3						
Power supply removal	1						
Remove bridges	1						
Push back	2						

A combination is also possible, e.g. a bridge attached to the aircraft door behind the crew cabin and mobile stairs for the rear exit. The aircraft doorsill height varies considerably depending on the aircraft type (Table 8–3). Therefore, it is sometimes necessary to have a range of passenger stairs or a range of stands equipped with different bridges at the ramp.

Although passenger bridges are more costly than mobile stairs, the former are increasingly used also at medium-sized airports. One of their advantages is that passengers can change between aircraft more quickly. While the aircraft is being boarded or deplaned, other servicing activities can be carried out simultaneously. The movement of the servicing vehicles across the apron is not obstructed. The safety of passengers is also ensured as the contact of passengers with servicing vehicles is avoided.

Table 8-3   **Door sill heights of some aircraft types**

Aircraft type	Door sill height [mm]
DC - 92	2130
Jak - 42	2600
B - 727	2670
Tu - 154	3200
DC - 8	3200
IL - 62	4200
A - 300	4500
B - 747	4700
DC - 10	4800

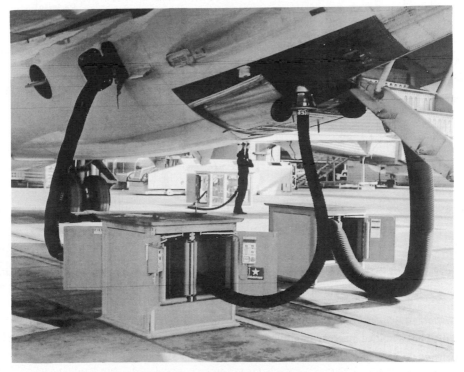

Figure 8-3   Fixed ground handling systems 'vehicle free apron' - FMT Pop-Up Ground
Support System (courtesy FMT - Fabriksmontering i Trelleborg AB)

The flow path of the passengers is straight-forward, and they reach their aircraft safely without getting lost. The quality of the whole process is improved as passengers are protected against bad weather. No significant change in level is experienced at the majority of airports and for most types of aircraft, as there are maximum allowable slopes on the bridge ramps. At some airports the passenger bridges are also used at the aircraft rear door, by extending over the aircraft wing. This speeds up boarding and deplaning and helps to shorten the turnround time.

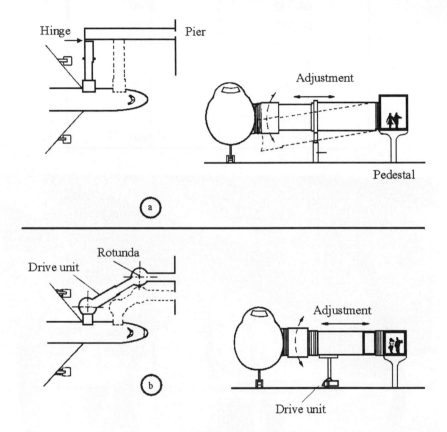

Figure 8-4   ***Types of passenger boarding bridges: a/ with a small range of adjustment and b/ with a wide range of adjustment***

However, passenger bridges are more costly than mobile stairs. Also if the passenger bridge installed has a limited manoeuvrability, the stand has less flexibility. Bridges of this type are less expensive than bridges which can be fully adjusted for all aircraft types.

Passenger boarding bridges with two telescopically connected parts consist of the following main parts:

✈ rotunda

✈ telescopic tunnel

✈ pedestal with undercarriage

✈ crew cab

✈ technological equipment.

The bridge can usually be extended by about 10 metres and the height can be adjusted within a range of almost 4 meters. The maximum incline of a bridge does not exceed 10%.

Figure 8-5   A passenger boarding bridge with a wide range of adjustment (photo A. Kazda)

8.2.2   Supplies of Power, Air-Conditioning and Compressed Air

Most aircraft can meet their energy requirements at a ramp with an APU. Advantages of such a solution are independence of the ground source and saving on time required to connect/disconnect the ground source. Among disadvantages are low efficiency of the APU (only about 30%), and high noise and environmental pollution due to exhaust gases. At Copenhagen airport APUs have to be turned off within 5 minutes after the plane docks and at Zürich airport this time period is reduced to 30 seconds. After that, the aircraft has to be supplied with electrical power of 400 Hz, 115/200 V. The distribution is provided either from the central power source or from a static (non-rotating) converter installed on the loading bridge. Banning APU usage at airports can reduce airport pollution by almost 10 %.

If there is no ground support system supplying air-conditioning, the pilot cannot avoid using APU in order to provide ventilating, heating or cooling in the cabin. In this case a ground electric source is not usually requested. Before the start of a long haul rotation, the aircraft needs power so that the maintenance engineers can check the systems. The seat-back facilities now being installed in aircraft require a lot of power and maintenance.

Air-conditioning has to be provided even in mild latitudes. Due to the high concentration of people aboard, cooling of the cabin has to be provided when the temperature rises above 10 °C. When the temperature is between 4.4 – 10 °C air has to be circulated, and when it drops below 4.4 °C the cabin has to be heated. Central distribution systems can supply air of - 6.5 °C in summer and + 66 °C in winter. In order to start the engines compressed air of 230 °C is supplied.

8.2.3   Cargo and Baggage Loading

Cargo and baggage loading is a critical activity especially when handling wide-bodied aircraft serving long-haul routes and on charter flights. When servicing the B-747, it is necessary to load and unload about 6 500 kg of baggage. An advantage of modern wide-bodied aircraft is that baggage is transported in containers. This reduces a number of bags going astray as the baggage having one destination is stored in one container. Handling loose cargo and baggage is the most serious problem because such handling is time-consuming. Some airports use transporters to handle them, or they build systems of conveyers to transport the baggage from the departure hall directly to the aircraft.

8.2.4   Push Back Operations

In order to improve apron utilisation, many international airports are equipped with nose-in aircraft stands. When ready for taxiing, the aircraft has to be pushed back into a position where it can use its own engines. Aircraft tugs of different designs are being used to push back the aircraft. The original push back unit was a traditional combination of a tractor and a towbar. An advantage of such a unit is a relatively lower initial cost when compared with other modern equipment. A disadvantage, especially at major airports, is the need for keeping different towbars for different types of aircraft. The towbar costs increase the overall investment needed for handling the nose-in stands. One towbar tractor requires at least two people to operate.

When compared with some other methods, this conventional model has a higher labour requirement. Other disadvantages, from an operational point of view, are low speeds of operational towing, e.g. between apron and a hangar, and problems arising from pushing back heavier aircraft under poor braking conditions on the apron. In order to provide the necessary adhesion, the tractors designed to handle the heaviest aircraft must also be very heavy. This results in a higher fuel consumption. At some airports electrically driven tractors are being used. At present there are several types of towbarless tractors designed to push back or tow aircraft.

Figure 8-6   A towbar tractor pushing back an aircraft (photo A. Kazda)

Powerpush from Finland is one company that has developed a push back vehicle on a fundamentally different basis. The Powerpush is towbarless. Unlike other vehicles for moving aircraft, this unit connects to the main landing gear, rather than to the nose wheel of the aircraft. Profiled rollers are placed hydraulically in contact with the tyres of the main gear wheels. The rollers are driven by a hydrostatic engine and make the wheels of the main gear rotate. The pilot controls the aircraft according to directives received from the Powerpush operator. The aircraft's brakes can be used without damaging either the aircraft or the push back vehicle. When compared with conventional vehicles, this unit is very light, simple, and both its operational and initial costs are lower. It can be remotely controlled. In such a case the operator of the vehicle has a direct visual contact with the pilot of the aircraft. Its disadvantage is that it can be used only for push back operations on an apron over a distance of from 50 to 150 m, and it is designed only for aircraft with one row of wheels on the main landing gear. The airport therefore needs to have other vehicles designed for operational towing. A new Powerpush version capable of handling wide-bodied aircraft is being developed currently.

The FMT Company in Sweden has presented another interesting version called Push Back System. It is partly built into the surface of the apron. On taxiing, the aircraft nosewheel falls into the 'cradle' of the equipment. On push back, the cradle is pulled by a hydraulically driven chain located under the surface of the apron.

Figure 8-7   Douglas TBL 400 Towbarless tractor (photo courtesy Douglas)

Many companies currently offer towbarless tractors designed for push back or operational towing at high speeds. In most cases the tractor picks up and lifts the aircraft's front wheel off the ground hydraulically. The wheel is clamped firmly. When compared with a conventional type, this tractor can be lighter as the required adhesion force is induced by the aircraft itself. The impacts which are transmitted to the nose undercarriage leg when a towbar is used for towing or pushing back, are, in this design, damped by the aircraft tyre. The main benefit of this system would come when towing an aircraft from the apron to the end of the runway before take off. The economic benefits are fuel savings on taxiing, e.g. a B-747 at Frankfurt needs on average 607 kg of fuel for taxiing to end of the runway. Consequently, there is a cleaner environment at the airport and its surroundings. For countries with high labour costs it is also of interest that the tractors of this design are one-man operated. The disadvantage is a high initial cost of the unit. In fact, due to persistent operational and legislative problems, it has not yet been possible to introduce operational towing. The design of the aircraft's nose gear legs have not been designed to transfer loads when towing an aircraft at higher speeds for long distances, but only for short distances on push back. It would be necessary to build new special tow tractor roads for the return journeys to apron from the holding bays. There is also resistance from the airlines, who fear that any problem with starting the engines could mean a long return trip to the stand. Another question which has not been decided is who is legally in control of an aircraft when it is under operational tow. Is it the pilot or the driver? Despite of all the questions mentioned so far, the costs of the vehicles will gradually go down and they can be used in major airports or the airports where a clean environment is of primary concern.

8.3 VISUAL GUIDANCE SYSTEMS

Semi-fixed passenger loading systems with a small range of adjustment are less expensive than the fully mobile bridges. They do, however, have less flexibility and need precise positioning of the aircraft in order to avoid damage and also to attach the equipment needed for the aircraft servicing.

It is, therefore, particularly necessary to install visual guidance systems to facilitate the positioning, though guidance systems are usually installed also at stands equipped with fully mobile bridges. Exact positioning is possible also when the aircraft is guided manually by a marshaller, but when an airport accommodates wide-bodied aircraft, the pilot often cannot see the marshaller easily from the cockpit. When the aircraft is damaged, it is often difficult to decide whether it is the marshaller or the pilot who is to be blamed for the damage. At large airports where passenger boarding bridges are usually installed it is therefore vital to implement guidance systems.

One of the first guidance systems was installed at the airport in Atlanta, Georgia in 1956. The simple and cheap equipment consisted of a stick attached on the passenger loading bridge to a tennis ball with a rope. It was installed at stands designed for the Douglas DC-6. The aircraft was parked correctly when the tennis ball touched the windscreen. A disadvantage was that the guidance system could be used for one aircraft type only or it would be necessary to move the tennis ball.

A modern but more advanced modification of the above mentioned stick, rope and tennis ball is the UCRAFT system developed by the French company UGEC S.A. Data about types of aircraft using the stand are stored in a computer memory. According to a pre-selected aircraft type the computer adjusts the position of a hydraulically controlled arm. At the end of the arm there is a light tube containing sensors. The aircraft has to touch the tube with its windscreen. If the aircraft does not stop, the arm keeps moving with the aircraft. However, it is expensive, at about GBP 60 000.

The Swedish company Inogon International AG makes use of two indicators working on a principle of interference of light to guide the aircraft along the centre-line and to indicate distance. In a box-like indicator there is a fluorescent light below which two grids are located and where interference of light takes place. The resultant patterns are called moiré.

Their shape depends on the line of vision. The centreline beacon provides directional guidance for the aircraft. Another indicator located obliquely from the pilot's line of vision gives the distance of the aircraft from the stop position at the stand. Black lines shaped like arrows direct a pilot to the right position on stand. In the stop position the pilot can see black perpendicular lines in the centre of both indicators. The system has no moving parts, its production and operation are simple. Another modern system installed at Arlanda, Stockholm is a system which uses data from the radar. The main advantage of this system called APIS (Aircraft Parking and Information System) is that it requires no sensors located on the tarmac. The radar measures the distance of the aircraft to its final position where it is to stop. The pilot keeps direction by following the indicator which makes use of interference of light on a grid, as already mentioned.

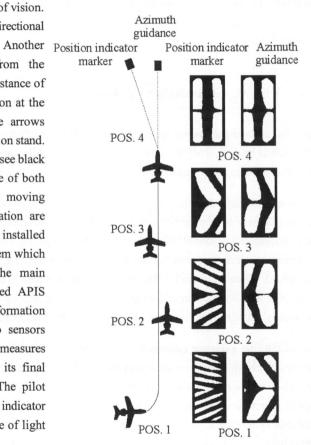

Exact parking of the aircraft is enabled also by another parking system produced by Safegate SATT Electronics AB. Its

Figure 8-8   *Visual guidance system of Swedish company Inogon working on the principle of interference of light*

advantage is that information about direction and distance is provided from a single source.

Centreline guidance is provided by a vertical illuminated green bar which is, from the pilot's line of vision, displayed in a yellow symbol of aircraft painted on the surface of the guidance panel. The information about forward motion of the aircraft is presented on a vertical panel through sequential activation of green lights. Exact, final positioning of the aircraft is indicated by flashing red lights. The position of the aircraft nosewheel is sensed through induction loops placed under the apron surface. The data concerning the aircraft position are processed by means of a microprocessor. Before the manoeuvre starts, the data about the aircraft type have to be recalled from the memory of the computer. They are, for the sake of checking, displayed in the top part of the guidance panel.

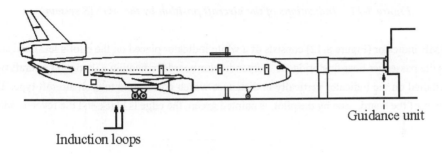

Guidance unit

Induction loops

Figure 8-9   Principle of Safegate visual guidance system

Most airports today prefer relatively low cost and simple systems. The airport authorities are less keen to use eqipment requiring sensors and cables to be installed on the tarmac. When new types of aircraft are later introduced and, consequently, the stand configuration is to be changed, relocation of sensors is a costly procedure.

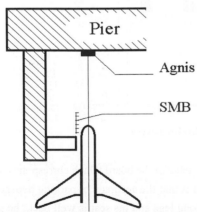

Pier

Agnis

SMB

Among these low cost, simple and reliable systems is one called AGNIS (Azimuth Guidance for Nose-in Stands) which are widely used at many airports for directional guidance of aircraft to stands, with a Side Marker Board (SMB) or Parallax Aircraft Parking Aid (PAPA) to indicate distance of the aircraft from its final position. Any skilled carpenter or electrician is able toconstruct SMB and PAPA.

The AGNIS indicator is placed on the stand centreline. Two vertical illuminated light tubes emit green and/or red light through openings, thus indicating a position of the aircraft on or away from

Figure 8-10   Placing of the AGNI and SMB systems

the parking centreline (Figures 8-10 and 8-11).

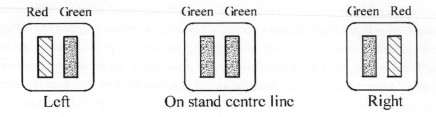

Red   Green   Green   Green   Green   Red

Left   On stand centre line   Right

Figure 8-11   Indications of the aircraft position by the AGNIS system

The SMB indicator (Figure 8-12) consists of a white indicator placed on the pilot's left-hand side, where the passenger boarding bridge is. Perpendicular boards indicating a particular aircraft type are installed on the indicator vertically in distances which correspond to each aircraft type. The front part of the board, seen by the pilot, is painted green, the edge is black and the reverse side is red.

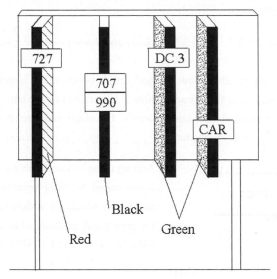

Figure 8-12   Side Marker Board

The pilot has to stop so that he can see only the black edge of the board which corresponds to the particular aircraft type. A disadvantage of the system is that the accuracy of parking depends on the position of the pilot's head. When the pilot has long legs and his seat is well back, he parks 'long' and the pilot with short legs parks 'short'. Some cases were reported with the B-737 when the pilot was too short a stall speed indicator placed near the aircraft door was damaged.

The PAPA indicator is located in the right side of the stand centreline. The indicator consists of a black board with a longitudinal opening. Aircraft positions are indicated through white perpendicular lines showing a particular aircraft type. At a certain distance behind the board there is a fluorescent light tube placed vertically. When taxiing to the stand the fluorescent light in the opening seems to move. When the aircraft is parked accurately, the pilot can see the fluorescent light tube behind the white line which corresponds to the particular aircraft type.

Figure 8-13   Side marker board (photo A. Kazda)

The principle on which the Burroughs Optical Lens Docking System (BOLDS) system operates, used mostly in the USA, is similar to that of the PAPA system. A difference is that the distance is indicated by means of a tube placed horizontally, and the direction through a vertical fluorescent light. Optics consisting of a Fresnel lens are placed in front of the light source. An advantage of the BOLDS, shared with the Safegate Docking System and UCRAFT, is that it provides information on both direction and distance from one source. When compared with the latter systems, the former is simpler and cheaper.

Also simple and low cost is the equipment installed at Gatwick in London for exact positioning of aircraft. Positions of the nosewheels of different aircraft types are painted on the apron surface. In front of the stand, on the left from the pilot's line of vision, there is a mirror in which the pilot can see the nosewheel of the aircraft when parking to a desired position. In winter the mirror is heated.

This survey of visual guidance systems is not complete. The choice of a suitable system depends on many factors, mainly traffic density, number of aircraft types using the particular stand and also on prevailing meteorological conditions. Future development of guidance systems will follow two paths which can be called respectively 'hi-tech' and 'beauty is in simplicity'. Meanwhile, the proliferation of types of guidance system can cause problems for pilots due to lack of consistency.

9

AIRCRAFT REFUELLING

Tony Kazda and Bob Caves

Motto: Fuel is the blood of aviation

9.1   BACKGROUND

In the early days of aviation, storing of fuel and aircraft refuelling was an undemanding activity. Piston engines had low consumption and the aeroplane was not able to carry a large quantity of fuel. The aeroplanes were refuelled from barrels or tankers. It was not necessary to build fuel stores at the airport. The same fuel as for cars was used for aeroplanes.

The requirements for quantity and quality of fuel gradually increased. Introduction of the jet aircraft in the late 1950´s changed overnight the requirements, both qualitatively and quantitatively, for storing and supplying the aircraft with fuels. New aeroplanes required greater quantities of fuel, yet had to be refuelled within the normal turnround time of less than 1 hour. Another great change occurred in 1970, when the wide–body Boeing 747 was introduced into operation. About 150 000 litres of fuel had to be transported through four refuelling openings with a diameter of 64 mm into the tanks of that aeroplane at a height of 4.5 m within approximately 45 minutes. If there was no hydrant fuel system at the airport, refuelling was provided by tankers with a capacity of up to 80 000 litres of fuel. At present the aircraft kerosene Jet A1 is the most common kind of fuel. For comparison, the following table gives volumes of the tanks of the selected types of aircraft:

Table 9-1   Fuel tanks capacity of selected types of airplanes

Airplane	Fuel tank capacity (standard) [1]	Fuel tank capacity (long range version) [1]
Prop- and turboprop		
IL – 14	3 500	
DC – 7C	29 800	
IL - 18	23 500	
Jet		
BAe 146	11 728	12 901
Caravelle	19 000	
B 737-500	20 105	23 170
A 300-600 R	68 150	73 000
IL-62 MK	105 000	
A 340-200	135 000	
DC 10-30	138 294	153 434
B 747-100B	181 970	
B 747-400	204 350	216 850

The consumption of fuel does not rise so rapidly as the air transport traffic. In the period 1960 to 1995, passenger km grew at 8.9 % per annum and aircraft departures grew at 2.7 % per annum. Over the same period, the fuel used per passenger km reduced by 2.5 % per annum. This improvement in fuel productivity is caused by:

✈ a gradual growth of the use of seat capacity, or passenger load factor, of aeroplanes, by the European airlines in particular

✈ a gradual substitution of new modern types of aeroplanes with lower unit consumption than the old ones

✈ changes of flight procedures and wide use of flight management systems.

Fuel productivity gains have slowed to about half these levels since the mid 1980s because the average aircraft size has stopped increasing and there has been more congestion in the system.

9.2   FUEL - REQUIREMENTS

The basic requirements for the provision of fuel at an airport are to ensure:

✈ high purity of fuel

✈ sufficient supply of fuel

✈ economic and rapid supply of fuel to aeroplanes with high safety standards

✈ environment protection.

9.2.1   Requirements for Fuel Quality

In order that the stringent safety requirements can be observed, the international organisations in co-operation with the major oil companies have prescribed criteria that the aircraft kerosene Jet A1 must comply with.

During the flight of an aeroplane at high altitudes, the temperature in the aircraft tanks drops under $-40\ °C$. Therefore the fuel must not freeze even during long time exposure to very low temperatures and must have the required viscosity. A resistance against freezing depends in particular on the proportion of paraffin in the fuel, which crystallises at low temperatures. With Jet A1, crystals of paraffin may not occur in the fuel above a temperature of $-47\ °C$.

Fuel is used in the aeroplane for cooling the on board equipment, and therefore it may be exposed to high temperatures. Even at a temperature of $260\ °C$, the fuel must be thermally stable and must not create deposits. Thermal instability of fuel may be caused by the presence of trace elements. Their precipitation is obtained by the use of additives to the fuel.

Fuel must not be corrosive so that it cannot damage parts of the fuel system and the aircraft engines. Corrosion is caused mostly by the presence of sulphur in the fuel. Sulphur is removed from fuel during its production. Some compounds of sulphur, for example hydrogen (mono)sulphide, may be created by the activity of micro-organisms.

Fuel must not contain active surfactants. The active surfactants act on the water as emulsifiers, and prevent its removal from the fuel. The active surfactants are removed from the fuel by means of clay filters which are located before the micro filters.

The fuel must not contain more than $0.003\ \%$ of free water. With higher contents of free water, the ice crystals might block the filters after a drop in temperature. In addition, the presence of water supports the growth of micro-organisms in fuel. The free and dissolved water is removed from the fuel during its production in refineries. It leaves the refinery within its specification. However, during its transport and storage, water gets into it by the so called 'breathing' of the storage tanks. 'Big breathing' indicates a penetration of fresh air into the storage tank while it is being emptied. The atmospheric humidity condenses when there is a variation of temperature on the walls of the storage tank, and water gets into the fuel. 'Small breathing' occurs in the storage tanks during a change in the volume of fuel and air, as a consequence of variations of temperature between day and night. In happens in the railway containers in particular. In addition, aircraft kerosene is hygroscopic, and in a damp atmosphere it readily absorbs water. Therefore water should be removed from the fuel every time it is transferred from one container to another.

Similarly, solid impurities get into fuel during its transport and storage. They are mostly rust and dust from the walls of the containers. As it enters the aircraft's tanks, the fuel must not contain more that 3 g of solid impurities per 1000 kg of fuel with the size of impurities not greater than 5 μm.

Large volumes of aircraft kerosene must be pumped into the tanks of an aeroplane within a relatively short time while ensuring its high, almost pharmaceutical quality. This is possible only with correct construction and maintenance of the equipment and by following the prescribed procedures.

The construction of storage tanks must eliminate any possibility of water or impurities getting into the storage tanks. The bottom of the storage tank must be given a gradient, and at the lowest point there must be a drain to draw off the water and sediment. Similarly, the pipework must also have a minimum gradient and at its lowest point equipped with sediment discharge valves.

Treatment of the fuel during its transport and storage must ensure that, when used, the quality of fuel approaches the same quality it had when it was leaving the distillation column. The delivery of fuel must be accompanied by a certificate guaranteeing its quality. The certificate should be presented whenever the fuel is handled. Techniques to maintain the quality include sedimentation, discharge of the water and sediment, suction from the surface, filtration, and separation.

After delivering the fuel into storage tanks in the airport, the fuel must rest quiescent for a prescribed period of time in order that the impurities and free water can settle to the bottom of the storage tank. The time for sedimentation of the impurities depends on the height of the fuel column in the storage tank, and is about 3 hours per metre. Therefore, in the airport there should be at least three storage tanks so that, at any time, one of them is being filled, another resting quiescent, and aircraft refuelling is done from the third.

Sucting the fuel from the surface ensures that the relatively cleanest fuel is drawn out. A suction pipe is divided into a fixed part and a movable part which are connected with a joint. In the end of the movable part there is a float which maintains the intake of the suction pipe just below the surface.

Filtration is a process by means of which solid impurities are removed from fuel. The filters operate with an efficiency of 3-10 micrometers. If the filter has a filtration efficiency of, for example, 5 micrometers, it does not mean that larger particles do not pass through the filter. Elongated impurities may pass through the filtering pore even if their length is more that 5 micrometers. In the majority of cases centrifugal separators are used for water separation, or a double-stage filter.

Sometimes, mostly in subtropical and tropical zones, micro-organisms may be observed in the fuel system filters or tanks. In the majority of cases, they are micro-organisms which use the aircraft kerosene as a power source for their metabolism. In some fuel stores *Cladosporium resinae* mould was discovered That species of micro-organisms is very small and easily passes even through fine filters. The protection consists in an exhaustive cleaning and disinfecting of the affected sections of the fuel system including replacement of filters.

9.2.2   Fuel Storage

The problem with ensuring the fuel stays within specification is that the demand varies considerably during the year. The consumption in the peak is approximately 60 % higher than in the off-peak months. In some airports there are also significant weekly variations in the demand. The storage capacities of fuel in the airport should be determined depending upon the reliability of fuel supplies and of variations in the demand. When determining the number of individual storage tanks and their capacities, consideration should be given to the peak daily consumption, the already mentioned reliability of fuel supplies, the state regulations and the required reserves of fuel.

The required reserve is the quantity of fuel that is in the store, that may not be used at that time to refuel the aeroplanes. Fuel must be treated (see the preceding section) and depending upon the manner of transport of the fuel to the airport, it is sometimes necessary to analyse it. In the event that the fuel samples do not correspond to the specification, the fuel may be returned to the supplier. Therefore the required reserve must cover a sufficient quantity so as to ensure a continuity of supply for airlines.

Fuel is transported to stores in the airport in the following ways:

✦ by a pipeline directly from the production plant or the oil company's storage facility

✦ by ships or barges

✦ by railway containers

✦ by lorry-mounted containers.

Transport by pipeline is particularly suitable for large airports, or if the airport is situated near a refinery. In such cases, and if a reliable supplementary supply of fuel is available, the requirement for constructing large stores of fuel directly on the airport is reduced. A pipeline supplies the fuel stores at Heathrow, Gatwick and Stansted airports. In some cases it is advantageous to transport the fuel by pipeline even for long distances. For example, a 100 km long pipeline supplies the Jet A1 from the refinery at São José dos Campos in Brasil to the new São Paulo Guarulhos airport.

Supplying fuel by truck is more suitable for small airports. However, there are exceptions. For example, Manchester airport was until recently supplied exclusively by truck, with about 100 containers per day.

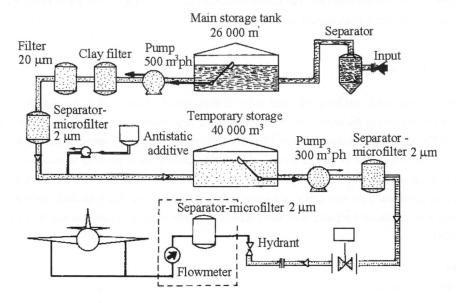

Figure 9-1   The fuel system at Charles de Gaulle airport

Figure 9-1 shows the fuel system at Charles de Gaulle airport in Paris. Before filling the fuel into the main storage tank, water is separated from the fuel in a separator. After sedimentation and the discharging of water and sediments from the storage tank, the fuel passes through a clay filter with an efficiency of 20 μm. Then the fuel passes through the fuel system. Static electricity accumulates in it due to the friction from the piping walls, so the system has been designed in such a manner that the velocity of fuel does not exceed 3 m.s^{-1}, and an antistatic additive is added. Static electricity is discharged from the fuel during its temporary storage in a delivery storage tank. Before aircraft refuelling, the fuel is once more cleaned in a filter with an efficiency of 2 μm and for the last time in a filter directly in the delivery dispenser.

Fuel storage is a potential danger which may affect the quality of underground and surface waters, and the soil environment. The danger may be limited by obeying the fuel technology and construction measures.

Storage tanks may be underground or overground, the decision being made on the basis of a special study for each case. The underground storage tanks are mostly used in smaller airports with volumes of individual storage tanks up to 120 m^3, rarely up to 300 m^3. The underground storage tanks are, with respect to national standards, usually designed with double jackets, or with a single jacket located in an impermeable caisson. When double-jacket storage tanks are used, the cost of the technology is higher, while when single-jacket ones are used the price of the construction works is higher. In some countries, a cathode protection of the storage tank against corrosion is considered to be sufficient to secure the storage tanks against fuel leakage. The price of the overground storage tanks themselves is usually lower than that of the underground ones. However, the fire regulations require the overground storage tanks to have greater safety distances from other objects and prescribe the availability of considerable fire-fighting equipment. The overground storage tanks are situated in pits which are able to retain the fuel in the event of an accident. The storage tanks should be cleaned on a regular basis, usually once in three years, and are subject to prescribed tests.

The fuel storage facilities at Prague Ruzyně airport have 6 storage tanks and a total volume of 4.8 million litres. Those at Miami International airport have a total of 32 storage tanks with a total volume of 72 million litres of fuel. An average daily supply of fuel in that airport is more than 5.7 million litres.

9.3   FUEL DISTRIBUTION

Distribution of fuel from the fuel storage on the airports into aeroplanes may be executed basically in two ways. One is a distribution to aeroplanes by tanker, the other is a distribution by fixed systems directly into aeroplanes. Both of the systems have their advantages and disadvantages, and both of them, naturally to a different degree, are used at small as well as large airports.

In small airports with a large number of general aviation aeroplanes, it is often more appropriate to install a fixed fuel delivering system, to which the aeroplanes will taxi. Similarly, it is advantageous to install a hydrant system, for example at a heliport with more stands. In both cases, the principal advantage is a reduction of operating costs and an increase of safety and danger of contamination during aircraft refuelling. At smaller airports, the larger aeroplanes are usually refuelled by tanker, since at airports with low volumes of fuel deliveries, it is not economical or efficient to install a fixed system to each individual stand. Within a certain time, fuel in the systems downgrades, its quality is not ensured, and part of the fuel should be sucked out before aircraft refuelling.

At the present time, the administrations of most large international airports prefer hydrant fuel distribution. The decision in favour of fixed installations is not only made on cost grounds, but also for operational reasons.

Figure 9-2   Aircraft refuelling by fuel dispenser (photo A. Kazda)

At an airport with wide body Boeings 747 turnrounds, the fuel supplies to a single aeroplane usually amounts to 100 m^3 or 80 tons of fuel. With a hydrant distribution system, in the majority of cases a single worker with one dispenser, which is only a light vehicle, can perform the aircraft refuelling. The dispenser has equipment for reducing the operating pressure of fuel in the system, which reduces from 2 MPa, to a value of 0.35 MPa pressure of aircraft refuelling. Besides that, there are filters and fuel quantity measuring instruments, and a height adjustable manipulation platform. A safety valve, most often of the Dead Man's Handle type, is installed between the hydrant and the dispenser, which in the event of a failure closes the fuel intake. In comparison, two or three tankers with operators are required for refuelling a high-capacity aeroplane. The tankers obstruct a large space around the aeroplane and impede other activities. Another advantage of the hydrant system is in an increase in safety. There are no tankers on the apron with a large quantity of highly inflammable fuel, but only a dispenser with fuel in hoses from the hydrant to the aeroplane and in the filters.

Tankers and dispensers must be parked somewhere. The operator prefers to park and fill tankers in the closest proximity possible to the apron in order to increase their utilisation and reduce costs. But the areas near the apron are commercially the most valued, and with the large size of vehicles, the price of the lease will be high. If the parking lot is located further away, the tankers add to the traffic on the service roads and in the event of an accident, they increase a possibility of spillage and fire. In contrast, the main parking area for dispensers may be situated even outside the central part of the airport. Only those dispensers that are immediately needed will be on the apron. The other equipment of the hydraulic system, storage tanks, pumps, filters, etc., are generally located off the apron.

The peak demand at Heathrow airport amounts to 18 000 m^3 of fuel. In typical large hydrant system alone there can be as much as 4 000 tonnes in the system at any one time. The delivery rates through a hose can be in excess of 300 l/min – equivalent to filling an average car in less than one second.

Unless they are fully drivable, passenger loading bridges limit to some extent the flexibility in where the aircraft can park in relation to the hydrants, but the hydrants also limit the location of the aircraft due to the limitation on the length of the hose from the hydrant to the aircraft refuelling point. However, at the same time, this type of airbridge ensures that the location of the stand for a particular aircraft type will not be change frequently. If there is a need, the hydrant may be repositioned by several metres over night, if the work is well organised.

Another possibility is the so called 'free ramp concept', where a hydrant system is installed without any elements of it being under the apron. Refuelling is provided directly from delivery points built into the airport apron, which are mounted flush with the level of the apron when not being used. An example of such a solution is at Stockholm-Arlanda airport.

At airports with a density of operations up to 5 million passengers per annum, the decision to be made between tankers or a hydrant system is often not simple. The investment costs of constructing the hydrant system may differ even between different locations on the same airport, depending on the extent of the system and the possibility of construction in stages. Consideration should be given to the following:

+ length of the distribution piping

+ building works, including embedding of the distribution system under the apron, taxiways or runways

+ number of stands

+ pay-back period.

The investment costs of constructing a hydrant system are several times higher than those of a purchase of several tankers. Nevertheless, its operating costs are lower and its service life longer. When projecting and constructing a hydrant system, it is necessary to ensure a high quality of works and materials because the costs of repair, reconstruction, or removal of the consequences of an environmental accident under the apron may be several times greater than the increased costs during the construction.

Even though a hydrant system of fuel distribution has been constructed in the airport, it is necessary to have available several tankers in case of failure or maintenance of a part of the hydrant system, so that fuel can still be supplied to the aeroplanes in aircraft parking stands, in hangars (for engine test), and for drawing the fuel off from the aeroplanes if there is a need.

The hydrant system in the Miami International Airport has more than 400 delivery hydrants. The total length of piping exceeds 40 km, and approximately 3.7 million litres of fuel are constantly filled in the piping.

In summary, the main advantages and disadvantages of hydrant systems are:

Advantages:

+ better fire safety

+ better ecological protection (lower probability of fuel escape and spilling)

+ smaller area required during technical servicing of aircraft

+ smaller area required for parking of dispensers in the vicinity of the apron

+ lower operating costs at high volumes of fuel supply

+ lower demands for labour

+ easier to ensure high quality of fuel

✈ aircraft may be very quickly Refuelled with any quantity of fuel

✈ lower demands for maintenance

✈ reduction of traffic on the service roads.

Disadvantages:

✈ higher investment costs

✈ the aeroplane should be quite accurately parked on the stand

✈ little flexibility

✈ high costs of repair in the event of a failure of the system

✈ need for several tankers to be available.

9.4   SAFETY OF THE REFUELLING OPERATION

Aircraft kerosene has a high concentration of harmful substances and high degree of flammability (class II). During the construction, operation and maintenance of a fuel system, adherence to the standards and set procedures should be ensured so that water cannot be polluted with fuel and so that humans and property cannot be harmed in the event of a fire.

9.4.1   Ecological Damage

Even small leaks, or spilling of fuel during refuelling may have the same serious consequences as a bigger escape of fuel after a failure of the equipment. Small escapes are more frequent, and more easily escape detection, but their effect is cumulative. Also, a small quantity of fuel contained in water can lead to significant levels of pollution, considerably lowering the quality of water and thus reducing its usability.

Fuel escapes may occur in three places:

✈ during transfer to or while stored in fuel stores

✈ from distribution piping

✈ during refuelling.

Then fuel may get into water by one of the following ways:

➢ by rain drainage into surface waters

➢ by sewage conduit into a water treatment station

➢ by infiltration into underground waters.

Fuel most often gets into the rain drainage after an escape of fuel on the apron. Aircraft kerosene is a poisonous substance and its most noticeable effect is in watercourses. After an escape of fuel, fish and other aquatic animals and aquatic plants are weakened by direct toxic effects of aircraft fuel and perish as a result of a deficiency of oxygen which is consumed as the microbes degrade the fuel.

Installation of a separator of oil products in the drainage branch from the apron, or in other parts of the drainage system, is a constructional measure taken to limit the fuel escapes through rain drainage channels. In some cases, the sewer discharges out at a retention tank serving among other things for retention of storm waters. In the event of an escape of fuel, the drain from the tank may be plugged, the layer of fuel may be withdrawn from the tank surface, and water may be treated before being discharged into a watercourse.

The simplest operational measure is to clean up fuel leaks during refuelling *in situ* by sprinkling them with an absorptive substance. After absorption, the residue is carried away and the fuel is burnt off in a specified place or in a special furnace. Special absorbant textiles, which are able to absorb up to 20 times their mass, are more suitable for removing the spilled fuel. It is simpler to remove the fuel with them than with loose material.

When greater quantities of fuel are spilled, some airports use special suction sweepers for removing the fuel. At Manchester airport, fuel is washed away with a large quantity of water into the sewerage drain, and later treated in a retention tank. The surface of the apron is thoroughly cleaned with a detergent. Detergent must not get into the sewage system, because the emulsion that would be created with water and fuel cannot be removed in the retention tank.

An efficient measure to reduce the number of cases of fuel spilling and to improve the discipline of handling agents providing aircraft refuelling is to invoice them with the costs connected with cleaning up.

Fuel escapes are often caused by the poor condition of the equipment. Therefore regular, visual inspection and scheduled maintenance are essential. Another measure is to prohibit execution of aircraft maintenance and repairs, except cases of removal of failed aircraft, on the apron, and permit them only in hangars, which are equipped with efficient separators.

Fuel may get into the sewage drains after leaks from aircraft in hangars and workshops. If collectors of oil products are not checked and cleaned on a regular basis, they cease to perform their function. If fuel gets into a water treatment station, it may disrupt the quality of the biological function of the station and disable it. Measures to prevent fuel escapes into a water treatment station are similar to those in the case of rain drainage.

When underground storage tanks or installations are damaged, or after a fuel escape on unsealed surfaces, fuel leaks into underground waters. Then it creates a layer on the water surface. New building regulations should prevent such an eventuality. However, there are still many installations that do not comply with the regulations. The solution then is for the implicated area to be surrounded by drill holes from whence the polluted water is continuously pumped out and treated. In addition, monitoring drill holes are established, from whence samples of water are taken. There is a disadvantage that by pumping, the level of underground waters drops, having an unfavourable impact on vegetation. Another solution is separation of the affected area by insertion of a plastic membrane at a depth of 30 m, or by partitioning off by ramming in steel walls. Then cleaning is performed only within the defined section. However, in any case, the state of the underground water is worsened

In 1983 approximately 1.1 million litres of fuel escaped from the hydrant distribution under the apron at Miami International Airport. They established 50 monitoring and 6 pumping drill holes through which they managed to remove 450 000 litres of aircraft kerosene from underground waters.

It is often difficult to detect a fuel leakage from the underground storage tanks or from the fuel piping. In spite of the fact that the majority of big airports have established a system of monitoring of fuel escapes by means of drill holes, it is only an identification of the consequences of the fuel escape. Marking substances may be utilised for an exact localisation of a failure and fuel escape. In the majority of cases, they are volatile materials which are added to the fuel. After several days, the place of damage location may be identified by means of a sensitive gas chromatograph. Since the marking substance has been used for locating the leak, the results are not distorted by any preceding pollution of land.

At some airports, the fuel system is periodically subjected to a pressure test. For example, at Frankfurt airport, the fuel system is taken out of service for one hour every night, and each section is separately pressured. A contingent drop of pressure is evaluated, and on the basis of the test results the maintenance of the hydrant system is executed. A disadvantage of this manner of testing is that a variation in temperature provokes a change in pressure, that may result in a false alarm.

Another manner of assessment of pipe leakage is used at the stage of performance tests. Very accurate flowmeters are installed in each section of the piping to measure the quantity of fuel at inlet and outlet from the respective section. The results are compared, and if there is a difference, an alarm is put out.

At each airport a plan should be elaborated for the event of an oil accident which all service personnel must be acquainted with. The plan shall contain not only procedures of catchment and removal of escaped material and removal of polluted earth, but also preventive measures to be taken to prevent an oil accident.

In some countries (e.g. USA), stringent limits of atmospheric emissions have been set by the regulations for environment protection. Among them, the contents of unburnt hydrocarbons have been regulated. The unburnt hydrocarbons may get into the atmosphere even during aircraft refuelling. Therefore on some airports, the vapours from aircraft tanks are drawn off and treated.

9.4.2   Fire Safety

A danger of fire during refuelling results from the combustibility of fuels. Aircraft gasoline is ranked among the class I inflammables (the highest rank) and aircraft kerosene into the class II inflammables. The prescribed provisions to protect against fire are related to that ranking. Among the passive ones is the extent of safety protective areas adjacent to the stores of aircraft fuels. Their size differs depending on the volume of storage tanks, their construction and the kind of stored material. The active measures prescribe the fire equipment, kind and quantity of extinguishing agents, procedures for fire fighting in compliance with the fire order in force.

Combustion of fuels may occur not only on contact with fire or with a material at high temperature, but also from a static electricity discharge. Therefore all parts of the fuel system must be provided with bonding and must be grounded. During aircraft refuelling, bonding should be provided also between the aircraft and the tanker or hydrant.

It is essential that the aircraft be fully restrained and that there should be a clear path for the bowsers to move away if an incident were to occur. Fuel zones should be established 6 m around the fuel vents. There should be emergency stop buttons on hydrant stands and no vehicle should be parked in front of it. There should also be an emergency phone at the head of the stand, and a person nominated as responsible for the fuelling operation. If fuelling is to occur while passengers are on board, the passengers should be told, the emergency exits should be kept available, and non-essential vehicles should be kept off the apron. If there is no airbridge, steps should be positioned both at front and rear doors, and the doors should be constantly manned.

Organising an intervention when a fuel escape occurs is usually within the competencies of fire brigades. They have both the required equipment and the required experience, they are able to assess the situation also from the viewpoint of the risks of possible fire.

9.5   AIRCRAFT FUEL - FUTURE TRENDS

It may be contemplated that aircraft kerosene will continue to be the most used aircraft fuel for at least the next 25 years. The assumptions that the crude oil reserves would soon be exhausted, which were made during the first oil crisis, proved to be unfounded. In the event that the crude oil reserves do eventually become exhausted, aircraft kerosene may be produced synthetically from other sources. The use of cryogenic fuels will be limited to supersonic aircraft with a speed over twice the speed of sound.

Thanks to improvements of the blade engine characteristics and other aspects of engine design, the consumption of aircraft kerosene will continue to increase more slowly than the air transport traffic. Such growth will alleviate problems in providing the quantities and speeds of fuel supplies into aircraft.

In reconstruction or construction of large airports, there will be a tendency prevailing to install modern hydrant systems. In small commercial airports, aeroplane refuelling from tankers will continue to be the most frequent solution.

A system of collection, transmission and processing of data on aeroplane refuelling is used at some airports. The monitoring system ensures a data collection and evaluation so that after refuelling a basis for invoicing may be available, or an invoice directly issued. Besides the fuel quantity, the modern computer equipment monitors also other parameters, such as fuel temperature, and automatically determines the total weight of fuel supplied into the aeroplane.

10

VISUAL AIDS FOR NAVIGATION

Tony Kazda, Bob Caves and František Bělohradský

10.1 MARKINGS

10.1.1 Markings Requirements

In the beginning of aviation the flights were performed only under visual meteorological conditions (VMC). At low speeds, the natural view of the terrain provided the pilot of an aeroplane with sufficient information for approach and landing under good meteorological conditions. As speeds increased, it was necessary to enhance the natural perception of the aerodrome by markings, and to provide the pilot of an aeroplane with additional information by marking the runway and other movement areas and to standardise the markings.

The terrain picture is enhanced by picture texture with objects, trees and other details. The picture texture is important particularly in the last phase of the final approach, during the actual landing and during take-off, when the pilot's attention is concentrated into a narrow visual angle in the direction of the runway and most of the texture is acquired by peripheral vision.

In the initial and middle approach phases and during a circuit, when the aeroplane is further from the aerodrome, the picture structure is important from the viewpoint of providing visual information. It is formed by the horizon and by natural features such as hills, valleys, rivers, fields, and the like. The pilot of an aeroplane may determine the aerodrome position by the picture structure.

Simple marking of ground movement areas is sufficient for conditions of impaired visibility by day, and in good meteorological conditions by night. Similarly, it is convenient to make use of the markings also for providing the pilot of an aeroplane with additional information during approach and landing, such as runway width and length. Natural properties of a picture seen by day in good visibility are complemented with these markings to emphasise and complement the crucial visual information on the location of the aerodrome and its runways. Only in extraordinary cases, when the aerodrome is located in an featureless landscape like desert, is it necessary to supplement the picture structure and texture more than this.

The requirements for visual information during a take-off are limited to emphasising the aircraft's position relative to the centre line and the start-of-roll end of the runway, and the ability to determine the speed and location of the aircraft relative to the take-off end of the runway during the take-off run, the rejection of a takeoff and after landing. During the final phase of an approach, the most important requirement is to be able to appreciate the runway width and length, and the runway length remaining after crossing the threshold.

The range of visual information required for safe movement of the aeroplane on the taxiway system and on the apron depends on the effective design of the taxiway system and apron. Directional guidance is given to the aeroplane to ensure that it stays on the taxiway and that it remains at a suitable distance from obstacles. The information necessary to facilitate the pilot's orientation at taxiway crossings and on the apron is again emphasised with markings.

In the final approach phase, the importance of the runway perspective, which is indicated by the converging angle of its edges, increases. The magnitude of the perspective angle is expressed by the relation:

$$tg\ \omega = \frac{d}{h}$$

where:

d   runway half-width

h   pilot's eye level above the runway.

During an approach, the perspective angle increases, and the pilot loses accurate information about the runway direction. Therefore on wider runways, it is necessary to mark not only the runway centre line, but to establish also the runway edge markings in order to improve the direction guidance of the aeroplane.

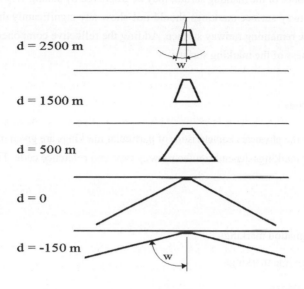

d - distance from the threshold

Figure 10-1   Change of perspective angle on approach with descent angle 2°30′
Source: J. Čihař, Letiště a jejich zařízení

The area of the runway system markings exceeds several thousands and even tens thousands of square metres. The choice of a material for markings should aim to minimise the total costs. The specifications of coating compositions for aerodrome markings differ from those of coating compositions for road markings. Especially at aerodromes with moderate traffic, the coating will be more affected by meteorological effects than by mechanical wearing. The paint for taxiway and apron markings should be colourfast. The durability of cheaper, chlorinated rubber paints is approximately 3 years, and machines may be used for marking. However, the total cost of renewal of the chlorinated rubber base paints may be higher in comparison with paints with longer durability.

Two-component paints are usually more expensive. They have a durability of 10 to 20 years, but, in general, it is more difficult to mechanise the marking. It is also necessary to consider that marking in a runway touchdown zone at aerodromes with a high traffic density will quickly become coated with rubber, and therefore it is useless to apply more expensive materials with longer durability.

The basic requirements for the markings are the colours, reflectance and roughness of the paint surface. The paint colours are determined by Annex 14 in the specified range of the colour spectrum. The reflectance of the marking surface may be improved by adding very small glass balls to the paint. The runway surface markings should not show any significantly different braking effect than that of the remaining runway surface. Adding the reflective component into the paint improves the roughness of the marking surface.

10.1.2 Marking Types

The requirements for the physical characteristics of particular markings are given in Annex 14. The extent of the runway markings depends on the runway type and reference code. They apply to the following markings:

Runway markings:

+ runway designation markings

+ runway centre line markings

+ threshold markings

+ aiming point markings

+ touch down zone markings

+ runway edge markings.

Taxiway markings:

+ taxiway centre line markings

+ taxi-holding position markings

+ taxiway intersection markings.

Apron markings:

+ aircraft stand marking

+ stand identification

+ lead-in line

+ turn bar

+ alignment bar

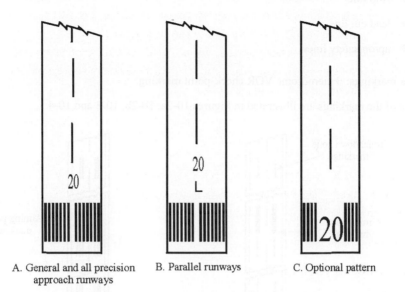

A. General and all precision approach runways

B. Parallel runways

C. Optional pattern

Figure 10-2a   Runway designation, centre line and threshold markings, Source Annex 14

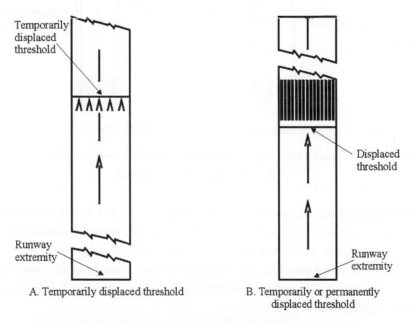

A. Temporarily displaced threshold

B. Temporarily or permanently displaced threshold

Figure 10-2b   Displaced threshold markings, Source Annex 14

+ stop line

+ lead out line

+ apron safety lines.

Other markings: + aerodrome VOR check-point marking.

Some of the markings are illustrated in Figures 10-2a, 10-2b, 10-3 and 10-4.

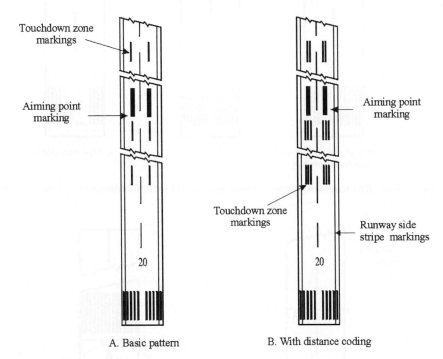

Figure 10-3   *Aiming point and touchdown zone markings*

Source: Annex 14, Aerodromes

An additional set of markings has recently been developed by the British Airports Authority to help pilots make better use of high speed exits. The markers are like those that announce the approach to exits on motorways, with three stripes at 300 m, two at 200 m and one with 100 m before the exit. The marker stripes have their own lights, of the same intensity as the runway lights. The benefit is that pilots maintain their speed after touchdown until the 300 m marker, so reducing runway occupancy time.

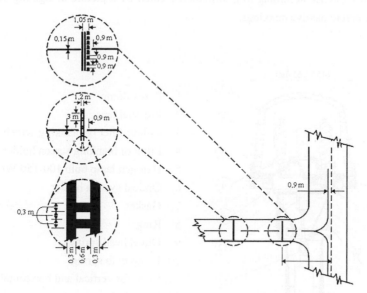

Figure 10-4   Taxiway and taxi-holding position marking – examples
Source: Annex 14, Aerodromes

10.2 AIRPORT LIGHTS   *(Author: František Bělohradský, Consultant, Prague, CZ)*

10.2.1 Characteristics of Airport Lights

10.2.1.1 Introduction

In the early days of aviation, the principal problem was to manage the flight itself. Later on, the aircraft performance improved. With the introduction of jet aircraft in the late 1950's and early 1960's, the aircraft parameters radically changed. At present, the civil aviation is oriented towards achieving high economic effectiveness with increasing regularity and safety of flight. This trend has been reflected not only in the design of airliners, but also in the area of airport design and ground equipment.

An important, and for the time being irreplaceable role is played, besides the radio-navigation and radio-location aids, also by the aerodrome lighting systems. They were all made more important by increases in approach speed, and the need to improve safety. In addition, the need for better reliability called for a reduction of meteorological limits to operation. Under conditions of impaired visibility during the day and at night, the last phase of approach and landing has to be performed with a visual reference according to information created by a picture of lighting systems that augment the purely passive markings.

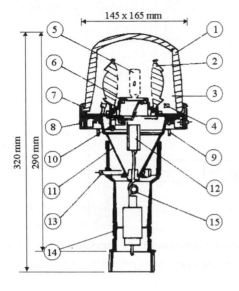

1. Pyrex dome
2. Fressnel lens
3. Colour filter or blanking screen
4. Filter or blanking screen holder
5. Halogen lamp bulb 100-150 W/ 6.6 A
6. Optical system holder
7. Gasket
8. Ring
9. Otical head spring
10. Lamp bulb spring
11. Cone for vertical and horizontal setting
12. Lamp bulb connector
13. Three setting bolts
14. Frangible coupling
15. FAA standard connector - plug

Figure 10-5   Omnidirectional light

As with other equipment, the aerodrome lighting systems are also standardised internationally by ICAO. The standards arose from the experience of Great Britain in particular, and later from the conclusions of work of the Visual Aids Panel (VAP) and All Weather Operation Panel (AWOP) teams.

The requirements for providing aerodromes with lighting systems vary with the type of runway equipment and reference code. The requirements are different for non-precision approach runways and for precision approach runways. Runways suitable for precision approaches are divided into three categories. Each category has a different meteorological limit for landing and take-off, and the requirements for lighting systems are also different. A survey of ICAO categories is given in the Table 10 - 1.

Table 10–1   ICAO precision approach categories

Parameter	Category				
	I	II	III A	III B	III C
Decision height	60 m (200 ft)	30 m (100 ft)	N	N	N
Visibility/RVR	800/550 m	- /350 m	- /200 m	- /50 m	N

Note: N – not defined

10.2.1.2 Approach and Runway Systems

10.2.1.2.1 Non-Instrument and Instrument Runways

Lighting systems of moderate light intensity are used for non-instrument runways under VMC conditions at night, and also for instrument runways light. It is often sufficient to construct only runway edge lights, runway threshold and end lights for non-instrument runways. If possible, it is advisable to construct a simple approach lighting system so as to improve the directional guidance. This is a firm requirement for instrument runways.

The simple approach lighting system consists of a single row of lights on the extended centre line of the runway, out to a distance of 420 m from the runway threshold with a crossbar at a distance of 300 m from the runway threshold (Figure 10-6). The simple approach system may consist of individual lights, or of short lighting bars, called barrettes. The effect of a barrette is created by at least three lights located next to each other. The runway edge lights and the threshold and runway end bars should be located at the runway edge, or at the most 3 m from its edge.

Omnidirectional lights are used for simple approach systems and edge lights for instrument runways and for non-instrument runways that are to be used under VMC at night. The omnidirectional lights provide guidance not only in the direction of approach and take-off, but because they give light in all directions, they also provide guidance for circling flight. Some types of lights combine the properties of reflector and omnidirectional lights. The combination of one omnidirectional light and two reflector lights that have been used for a precision approach runway has now been substituted by a single light unit. It represents an important capital cost saving.

The Thorn company's EL-EAH light is an example of such a light. The ICAO requirements for lights on runways with a width of 60 m are fulfilled with a power of only 150 W, while for runways with a width of 45 m, 100 W is a sufficient input with reserves of 20 to 50 % at a medium light intensity. The light (see Figure 10-5) weighs only 1.9 kg and has excellent aerodynamic properties. Resistance to wind and jet engine blast was tested in an wind tunnel, and it is well above the required 560 km.h^{-1}. Rust resistance is provided by the use of aluminium and stainless steel. The optical system is formed by a diopter of heat resistant glass and two Fresnel lens. The optical system bearer is formed by a solid aluminium casting, which in addition ensures good heat removal.

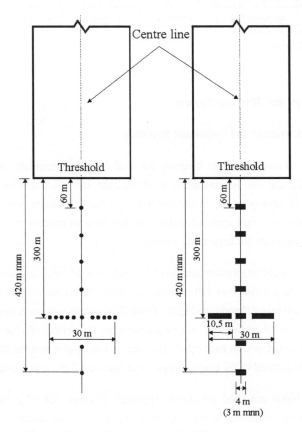

Figure 10-6   Simple approach lighting systems

Changing the lamp bulb is very easy and quick, which is particularly important at aerodromes with a high traffic density. The lamp bulb is equipped with a connector. The whole upper part of the light is usually changed for a new one, the lamp bulb being checked in the maintenance workshop. A change of the lamp bulb does not usually require more than 30 seconds. The mounting for the light is frangible (guidance on design for frangibility is contained in the Aerodrome Design Manual).

10.2.1.2.2 Precision Approach Runway

A Category I runway should be equipped with a Category I approach system. A diagram of the lighting system is shown in Figure 10-7. Two types of approach systems have been approved as a world-wide standard, an older type system, CALVERT and a newer type, ALPA ATA. The systems differ not only in the location of the lights, but also in the types used.

In addition, the ALPA ATA system has sequenced flashing lights, which more easily distinguishes the lighting system from other lighting in the aerodrome surroundings. The sequenced lights flash twice a second. The flash moves along the system towards the runway threshold. The time of each flash is so short that it does not dazzle the crew of an aircraft. The flash easily penetrates fog, and thus the pilot gets information about the runway centre line orientation several seconds earlier than from approach light systems with steady lighting.

A high voltage supply for the discharge lamps for the sequenced flashing lights is provided from a control box, which is fixed to the light support or is located in the ground.

ICAO states the requirements for the lights' properties, including the character of the white colour of the light and the mean light intensity in a defined beam of 20 000 candelas [cd] The requirements are usually complied with by reflector lights.

Threshold lights forming a crossbar should be located at the runway threshold (see Figure 10-8.). The threshold lights produce a green light with a light intensity in a defined beam away from the runway of 10 000 candelas [cd]. Runway end lights are also located at the runway end, facing back down the runway. The lights inform the pilot of an aeroplane of the runway end position. They produce red light in a defined beam with a light intensity of 2 500 candelas [cd].

The runway edges are indicated with runway edge lights. They are either reflector or omnidirectional lights producing permanent white light with a mean light intensity of 10 000 candelas [cd] in a defined beam. In the last 600 m of the runway length, the lights are yellow, and thus inform the pilot of the distance from the runway end.

The lights are supplied in the majority of cases by a serial distribution system and constant current regulator as described in Chapter 11. When the electric power supply from the mains is interrupted, the supply is provided from a standby power source. For a Category I runway, the electric power supply should be restored within fifteen seconds.

Category II. of the ICAO meteorological limits is characterised by a runway visual range from 400 m to 800 m and by a decision height of 30 m. These requirements are at present considered to be the limits for aircraft and crews to perform manual approaches and landings.

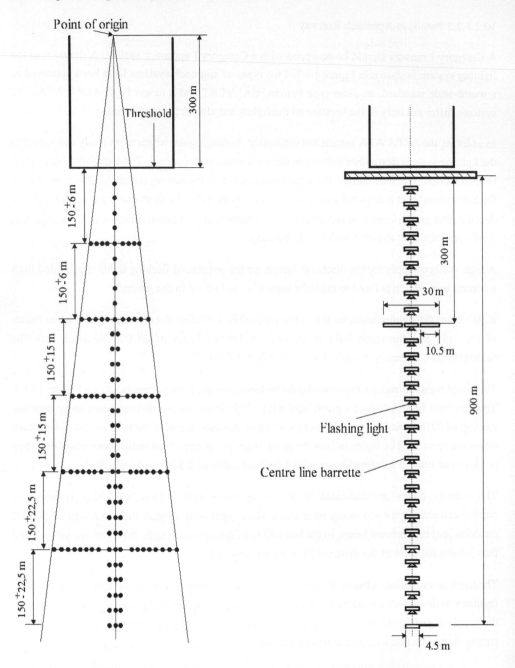

Figure 10-7   Precision approach Category I lighting systems
left –CALVERT, right ALPA-ATA

The Category I approach lighting system (CALVERT or ALPA ATA) is augmented for Category II operations with red side barrettes in the last 270 m before the runway threshold. The reflector lights have a light intensity of 5 000 candelas [cd] in a defined beam. The approach system also contains a set of flashing lights.

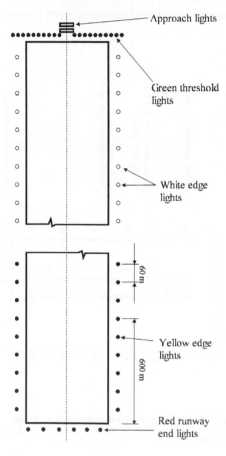

A Category II. runway has many more lights than a Category I. runway (see Figure 10-10). Besides the runway threshold lights, two other important additional lighting systems are incorporated. Under Category II meteorological conditions, the efficiency of the runway edge lights is reduced, and the runway centre line marked with inset lights becomes the most important means of giving precision guidance to a pilot on the runway during landing, as well as during take-off.

The centre lights show steady white light of variable intensity. Inset white and red lights shall be placed alternately in the centre line from the last 900 m to 300 m from the runway end, while the centre lights show only red in the last 300 m. Another system indicates the runway touchdown zone, again by means of inset lights. The lights of the touchdown zone shall be located with a longitudinal spacing equal to that of the runway edge lights, and form an extension of the series of approach barrettes. Contrary to the lights of the approach system barrettes, which are red, the touchdown zone lights are white

Figure 10-8   Runway lighting -
precision approach Category I runway

The inset lights are often run over by the wheels of aircraft. Their construction needs to allow for that, and they should be fitted flush with the pavement surface. Their egg-shaped body is made either of cast steel or of aluminium alloys. The lid is provided with cut-outs for the light beams. The optical system inside the light consists of a light source and reflector, glass prism and colour filter, as the case may be. The lights are stressed not only dynamically as the aircraft runs over them, but also thermally, by the heat of the lamp. Even in heavy traffic conditions, in damp and in aggressive environments like winter maintenance, they need to be reliable. A white light with a light intensity of 5 000 candelas [cd] is required in a defined beam. The most advanced inset lights use a special light source, whose halogen lamp and a reflector form a single integrated unit. The halogen lamps have a minimum size and long service life. Effective optical systems allow the power requirements of the lights to be reduced, so reducing the heat inside the lights, and thus extending the service life of the lamps. The design of the lights is simple, the operation and maintenance is much cheaper than those of the earlier types of lights.

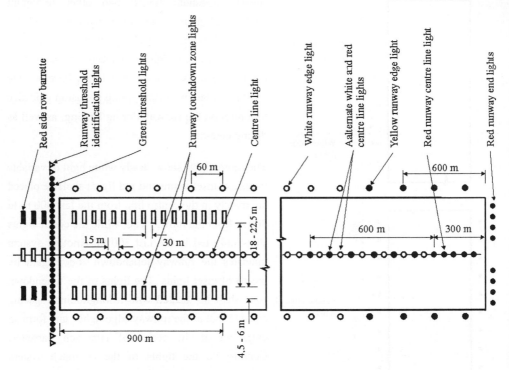

Figure 10-9   Inner 300 m approach and runway lighting for precision approach runway Categories II and III.

The IN-TO light made by the Thorn company (Figure 10-11) is an example of a modern inset light for taxiways. Optical prisms, which were glued in the older types of lights, are fixed into the light cap by means of a silicon seal. The light is provided with a double seal, the optical system being closed with a cover with a seal, and the other seal being between the cap and the mounting. The light power requirement is only 40 W.

The connection between the electric power supply from the isolating transformer and the light cap is waterproof. The colour filters for obtaining the required light colour are dichroic – a thin layer

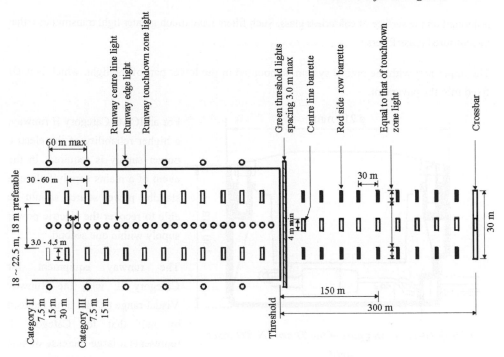

Figure 10-10 Runway lighting for precision approach runway Categories II and III

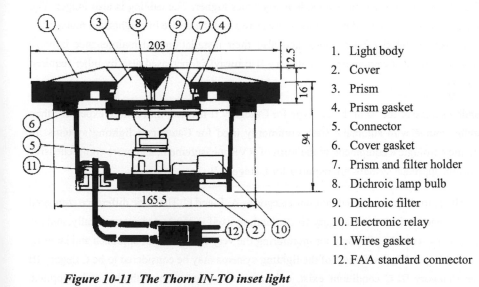

1. Light body
2. Cover
3. Prism
4. Prism gasket
5. Connector
6. Cover gasket
7. Prism and filter holder
8. Dichroic lamp bulb
9. Dichroic filter
10. Electronic relay
11. Wires gasket
12. FAA standard connector

Figure 10-11 The Thorn IN-TO inset light

is applied on the surface of colourless glass. Such filters have much greater light transmittance that the coloured glass filters.

The upper part with the optical system is mounted in the lower part of the light, which is itself fixed into the pavement.

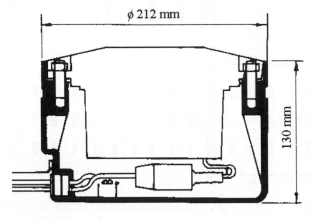

Figure 10-12 Lower part of the Thorn IN-TO inset light

For an ICAO Category II runway, a higher reliability of the electric power supply is required. In the event of a mains supply failure, a standby power source should be able to recover the electric power supply within one second.

The runway equipment for Category II, with the Runway Visual range (RVR) limit reduced by half that for Category I, represents a large increase of cost in comparison with a Category I runway. The number of lights and other components and equipment, such as isolating transformers, light fixings, constant current regulators, is many times higher. The cabling is also longer. The flush lights, which must be used in the touchdown rows and the centre line lights, are much more expensive than the usual elevated ones, and also their installation and maintenance is more expensive. The price of equipment for Category II runways is increased also by other required facilities.

The standby source of the 'short break' type for Category II is many times more expensive than the standby source with automatic start, commonly used for Category I lighting systems. In addition, meteorological equipment, in the form of RVR measurement, and more comprehensive ILS precision approach systems are required for Category II.

Category III approaches are divided into sub-categories A, B and C. The basic difference compared with other categories is that in Category III the landing itself is conducted automatically, and the lighting systems serve the pilot only for monitoring the very final phase of approach and landing. The practical limit of the usefulness of the lighting systems may be considered to be Category III B. When Category III C conditions exist, only non-visual approach methods can be applied. Landing and guidance of an aeroplane to the runway is fully automated.

The location of lights of the Category III approach and runway systems does not differ from Category II. The edge lights are not normally visible in Category III conditions, so the lighting systems of taxiways should be complemented with centre guidance from inset lights, which are green, or green and yellow from the runway centre line to the perimeter of the ILS/MLS critical or sensitive area.

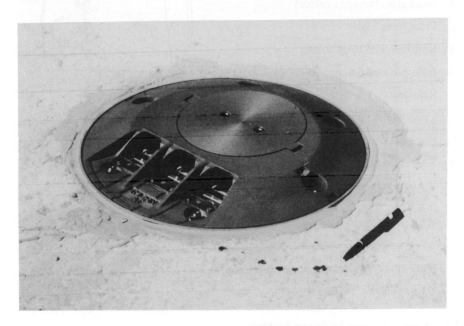

Figure 10-13 Photo of inset light – touch down zone light. A ball pen on the right is for size reference (photo A. Kazda)

For Category I, II and III approach systems, reflector type elevated lights are used. They (see Figure 10-14) should be light-weight, highly resistant against corrosion and jet blast. The reflector is usually made of aluminium, chemically polished and protected with a cover of light-weight aluminium alloy. The diffuser for green and red lights is made of heat resistant glass, and therefore no colour filter and its attachment are needed. The cooler in the rear part of the reflector lamp significantly extends its service life. Changing the lamp is facilitated by easy access to the lamp after lifting the rear plastic cover. The lamp is equipped with a connector. The power requirement is very small, being only 150 W for a white approach light, and 200 W for a threshold light, the ICAO requirements for light intensity still being fulfilled with reserves of 20 to 40 %.

1. Body in cast aluminium
2. Front glass (white or colour)
3. Front glass holder
4. Reflector with lamp holder
5. Halogen lamp 6.6. A, 150-200 W
6. Rear door
7. Lamp bulb holder
8. Frangible coupling
9. FAA standard connector - plug

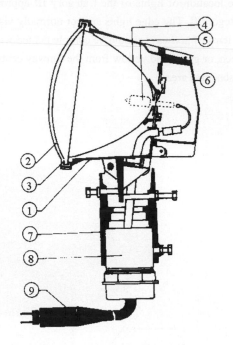

Figure 10-14 Reflector type elevated light for Cat. I, II and III approach systems

10.2.1.3 Approach Slope Indicator Systems

Where an improved visual guidance of an aeroplane should be provided in the approach slope, the runway must be equipped with a visual approach slope indicator system. The approach slope system should be provided if:

✈ the runway is used by jet aircraft

✈ the approach is to be performed over a water surface, or over a featureless terrain in the absence of sufficient incidental lights in the approach area by night

✈ misleading information is produced by surrounding terrain or runway slopes (see Figure 10-16)

✈ dangerous objects are present under the approach path

✈ the terrain beyond the runway ends involves serious hazard for an aeroplane if it touches down before the threshold or overruns beyond the runway end

✈ the aeroplane may be subjected to turbulence during approach.

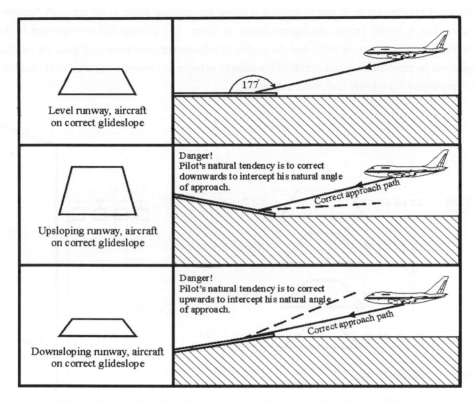

Level runway, aircraft
on correct glideslope

177

Danger!
Pilot's natural tendency is to correct
downwards to intercept his natural angle
of approach.

Correct approach path

Upsloping runway, aircraft
on correct glideslope

Danger!
Pilot's natural tendency is to correct
upwards to intercept his natural angle
of approach.

Correct approach path

Downsloping runway, aircraft
on correct glideslope

Figure 10-15   Possible illusions due to the longitudinal runway slopes

Visual approach slope indicator systems are installed at virtually every civil aerodrome. An additional advantage is that they correct any tendency to fly low in heavy rain. The most used is the Precision Approach Path Indicator (PAPI) system, and in some countries still the Visual Approach Slope Indicator System Visual Approach Slope Indicator System (VASIS), or T-VASIS in Australia and New Zealand. The PAPI system is not only a more precise vertical guidance of the pilot during the approach down the glideslope, but it also has fewer lights in the system. It gives a better indication of rates of change of angle above or below the glideslope. The costs of acquisition and construction are lower, and it is more economical to operate.

The PAPI visual slope light unit has a light beam horizontally divided into two sectors, red in the lower part and white in the upper part. The geometrical arrangement and horizontal adjustment of the lights ensure that the pilot of an aeroplane can see red sectors of the two lights nearest to the runway and white sectors of the other two lights when on the correct approach slope. If the aircraft falls below the glideslope, the white colour of the other two lights changes to red at precise angular changes depending on the extent to which it is below the nominal slope. If the approach becomes higher than it should be, the red lights change to white. The geometrical arrangement of the approach slope indicator systems and the angles of adjustment are shown in Figure 10-16. The lights can be repeated on the other side of the runway in order to improve roll perspective, but both sets must give completely consistent indications.

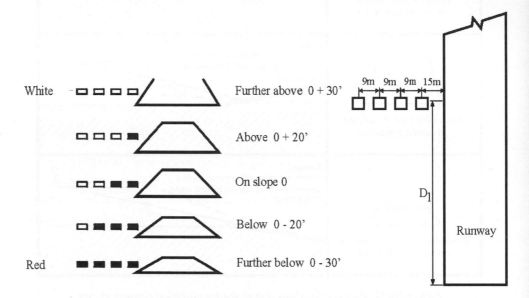

Figure 10-16 PAPI light units settings and indications

The construction of the visual slope light units does not resemble reflector or omnidirectional lights in any aspect. The light beam should be very sharply divided into a white and a red sector. The optical system contains one or two lenses, a red filter and a reflector.

It is normal to fit two optical systems in one light. The required power is then 2 x 200 W. In the APAPI systems – approved for aerodromes without international traffic, lamps of 45 and 100 W are also used, because the required range of the systems is lower. The required range of the PAPI system is 7.4 km on a clear day. The range is, in fact, usually 12 to 15 km, and, on a clear night, it can be 30 km and more.

The American Federal Aviation Administration (FAA) has approved another type of approach slope indicator system. It is the Pulse Light Approach Slope Indicator (PLASI) made by the De Vore Aviation Corporation. The equipment may be used on aerodromes as well as on heliports.

The equipment consists of a single light fitting. As with the VASIS and PAPI systems, PLASI also uses white and red lights to indicate the required vertical position of an aeroplane.

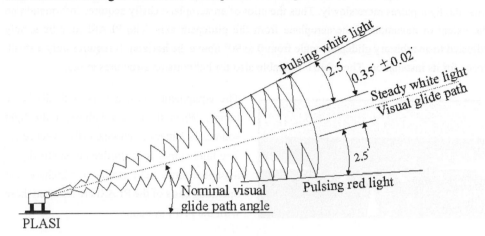

Figure 10-17 Principle of the Precision Light Approach Slope Indicator (PLASI)

A constant white light indicates that the aeroplane is correctly located relative to the glidepath. A pulsing white light indicates that the aeroplane is above, and the pulsing red light that it is under the glidepath.

Figure 10-18 T-VASIS light unit, Hamilton Airport, New Zealand (photo A. Kazda)

The transition between the position on the glidepath and off it is very sharp. The bigger the deviation of an aeroplane from the glidepath, the higher is the frequency of flashing. Nearer to the axis, the light pulses more slowly. Thus the pilot of an aeroplane easily acquires information on the extent of deviation of the aeroplane from the glidepath axis. The PLASI may be simply adjusted to an arbitrary glidepath angle from 0 to 90° above the horizon. It requires only a small. space for its installation. Therefore it is usable also for heliports on structures at sea.

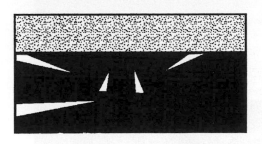

Figure 10-19 Principle of Glisada system
↗ *on slope*
↗ *left and above the slope*
↗ *right and below the slope*

The equipment has a battery with discharge lamps. When there is a failure of the light source, the battery is automatically switched on. The light output is controlled automatically by a photoelectric cell. Both the purchase and operating costs of the PLASI are less than those with the PAPI system.

In some states, tests have been performed using lasers, which could substitute not only the approach slope indication systems, but also the approach systems and the runway edge lights. The main advantage of a laser beam is that it is visible not only from a direct view into the light source, but thanks to scattering of aerosol particles and dust in the atmosphere, the 'materialised' beam is visible also beyond the axis of radiation. The visibility of the beam in a light air haze exceeds the natural visibility by up to six times.

The Glisada system used five red krypton and helium-neon lasers located along the runway, at the touch-down point and 60 m before the runway threshold. It was not used in civil operation for fear that the pilot's sight might be damaged.

For airports and heliports a new system of azimuth guidance was developed by the Thorn company. It is the System of Azimuth Guidance for Approach (SAGA). The system includes two lights with unidirectional rotating beams placed symmetrically on each side of the threshold.

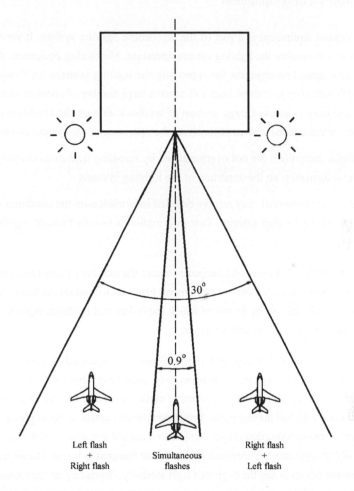

Figure 10-20 Principle of System of Approach Azimuth Visual Guidance (SAGA)

The pilot receives, every second, a luminous message comprising two flashes issued in sequence by the two rotating beams of the system. When the aircraft flies inside a 0.9 degree wide angular sector, centred on the approach axis, the pilot sees two lights flashing simultaneously. When the pilot flies inside a 30 degree wide angular sector, he sees the two lights flashing with a delay which varies between 60 and 330 ms according to the aircraft position in the sector. The further the plane is from the axis, the greater the delay, the sequence effect showing the direction to the axis.

10.2.2 Lighting System Components

10.2.2.1 Remote Control Equipment

The remote control equipment is a part of the aerodrome lighting system. It serves not only to regulate, but also to monitor the lighting system operation. Monitoring equipment that operates in real time is an essential requirement for approving the lighting systems for Category II and III conditions. It is necessary to control from a distance a large number of constant current regulators, and at the same time to assess a large amount of feedback data on the condition of the lighting systems. The remote control system rationalises the operation of the system so that:

→ the air traffic controllers are not overburdened by handling numerous control elements and redundant information on the condition of the lighting systems

→ the maintenance personnel may receive detailed information on the condition of the lighting systems, as a base for their activity. That information is usually brought together in a central control board.

The systems at small aerodromes and heliports, where the number of constant current regulators is small, do not require complex control equipment. From each constant current regulator, a multi-cored control cable is led out, by means of which the orders and feedback signals are transmitted directly to a simple board in the control tower.

At aerodromes equipped with Category I, II and III precision approach systems, a whole range of constant current regulators operates simultaneously, often to different degrees of light intensity. Movement of aircraft on more complex systems of taxiways is controlled by switching on the lights for the appropriate standard taxiing paths. Pre-programmed control systems prevent the air traffic controllers from being overburdened by a complicated control situation in such cases. For example, in the 'take-off' programme, the approach system is not required to be on. The air traffic controller chooses a runway direction and the degree of light intensity, depending on the runway visual range or according to the pilot's request. The programme ensures the activation of the lighting systems necessary for take-off with the required degrees of light intensity. In addition to control panels, clear diagrams are also used, where the activation of the particular lighting systems is shown. Lamps were initially used to show this, later being replaced by light emitting diodes (LED). The air traffic controller was informed of a failure of a part of the lighting systems or constant current regulators by the flickering of a lamp or a diode, and by aural signalling that can be switched off it necessary.

Current practice makes full use of computer technology for the control of the more complicated Category I systems, but in particular for Category II and III systems. This has both operating and economic advantages. The expensive multi-cored cable is substituted by a cheap two- or four-cored one. Large lighting systems and sub-systems, safety lighting bars and STOP lighting bars may be controlled and monitored, together with the power system of the aerodrome and meteorological and radio navigation equipment.

The air traffic controllers regulate the lighting systems through several controls, and the information is presented in a summarised form in monitors. For example, in the event of a failure of an important system, the air traffic controller may receive only the information that the lighting system is not capable of operation under the Category II conditions, but that it complies with the requirements for an operation under Category I conditions. For maintenance purposes, the monitoring system provides detailed data on the condition of the equipment failure in terms of reduction of insulation in a loop, the number of lamps not meeting the required output, and the like. In real time, the monitoring systems record the condition of the lighting systems, air traffic controllers activity, failures and their repair.

Recently, the so called 'addressing system' of monitoring and control has been used for complex light systems of taxiways and stop–bar lights. Two systems of communication are used, either a serial feeding cable, or a special cable network and optical cables.

Each light in a serial loop has its own precise address, and the system allows not only monitoring, e.g. it identifies two consecutive unlit lights, but also the control, individually or in groups of stop-bar lights, sections of runway centre line lights, etc.

The basis of the first system is the so called intelligent transformer. Between the transformer and the light source, the Micro Controller and Communication (MCC) switching unit is inserted. It contains the exact address of the light, and receives instructions for particular modes through a serial loop. The constant current regulator contains a Programmable Logic Controller (PLC) plate. Each loop has its own input into the control and monitoring system. An advantage of this system is that it may be installed in addition to an already existing system.

The other system provides many more possibilities of control by means of a specially created network. It utilises three types of intelligent units:

✈ Intelligent Light System (ILS) for lights

✈ Intelligent Detection System (IDS) for control and monitoring of the sensors at aircraft crossing points

✈ Intelligent Router (IR 2S or IR 3S) two or three channel system for network optimising.

The routers give shorter runs, and so make the cable network more economical. These systems are offered by virtually all the major manufacturers of lighting systems. They are used for complicated systems of taxiways for operations in Category III conditions at large aerodromes and in all conditions at aerodromes with high movement rates.

10.2.2.2 Light sources

The first generation of light sources were paraffin flares. These were supplanted by classical lamps designed for a maximum current value of 6.6 A with a bayonet cap. They are still used at many aerodromes for taxiway edge lights and to illuminate signs. They are made in a power input range of 30, 45, 65, 100, 200 and 300 W. Disadvantages are their large size and relatively small light flux.

First generation halogen lamps, of the so called double cap type, brought about a considerable reduction of the outer dimensions of lamps and an increase of light flux. Up to now they have been used in older types of inset lights. It was often necessary to refocus the lamp filament against the reflector, and the lamps have a relatively short service life because they were extremely heat stressed in the inset light fittings.

The second generation of lights, particularly the elevated ones, use halogen lamps with an exact position of the filament. They have extremely small outer dimensions, high light flux and a long service life of up to several thousand hours. They are made in an output range of 30, 45, 65, 100, 150 and 200 W.

Halogen lamps with an in-built miniature reflector in a single integrated unit are the most recent light sources to be applied in inset lights. These make it possible to design highly effective optical systems, and reduce the required input of the lights. The light sources made at present have inputs of 40, 45 and 100 W. They are outstanding for their long service life of up to 1 500 hours at 6.6 A., equivalent in normal use of up to 8 000 hours of service before failure.

The new generation of light sources improves the operating reliability of the inset lights, and at the same time reduces the costs of electrical energy and maintenance of the lighting systems.

10.2.2.3 Lights and Fittings

According to Annex 14, the aeronautical ground light is defined as:

Aeronautical ground light: any light specially provided as an aid to air navigation, other than a light displayed on an aircraft.'

The ground lights may be classified according to the following criteria:

✈ place of installation

✈ type of mounting

✈ radiation characteristics

✈ character of the light..

By where they are located in the aerodrome lighting system, the lights may be divided into:

✈ obstacle lights

✈ approach lights

✈ visual approach slope lights

✈ threshold lights

✈ runway centre line lights

✈ runway touch-down zone lights

✈ runway edge lights

✈ runway end lights

✈ taxiway edge lights

✈ taxiway centre lights

✈ STOP-bar lights

✈ safety lighting bars or 'wig-wags'

✈ lighting of information boards.

The types of mounting are:

✈ elevated

✈ inset.

The elevated lights are supported above ground level. The inset lights are firmly set in the pavement. Their cap protrudes only a few millimetres above the surface, and can bear the forces created when the wheels of an aeroplane or a ground vehicle roll over them. The optical system of the light is fixed in the cap.

The radiation characteristic of the lights may be divided into:

✈ reflector: inset lights may be unidirectional lights or bi-directional lights.

✈ omnidirectional lights, which produce the light into all angles of the azimuth.

The radiation characteristic of the lights may be:

✈ symmetrical

✈ asymmetrical

✈ semi-asymmetrical.

The character of the lights may be classified as:

✈ constant light

✈ flashing

✈ flickering.

Using these characteristics, it is possible to create names for the lights and thus specify them further, for example:

✈ elevated reflector approach light

✈ unidirectional inset runway centre line light

✈ elevated omnidirectional taxiway edge light

✈ bi-directional inset taxiway centre line light.

10.2.2.4 Requirements for Aerodrome Lights

The requirements for lights, their luminosity characteristics, mechanical properties and construction are determined not only by Annex 14 and other ICAO documents, but also by national standards. The requirements for lights may be divided into:

✈ lighting

✈ mechanical fitting

✈ electrical

✈ other requirements.

The lighting requirements are determined by Annex 14 by isocandela curves (Figure 10-21). Isocandela curves for the lights designed for Category I operations are wider horizontally. In contrast, the requirements for light intensity are higher vertically and lower horizontally for Category II. That requirement is even more conspicuous with the lights designed for Category III. All the manufacturers of lighting systems make efforts to provide lights that comply with the requirements for installation into systems for any of the Categories I to III.

Two basic factors are specified for evaluating the luminosity characteristics of the lights:

✈ mean light intensity, which is the integral of light intensity in a space beam (Istr)

✈ evenness of distribution of the light intensity: in a defined beam a minimum light intensity should be 0.5 Istr and a maximum light intensity of 1.5 Istr.

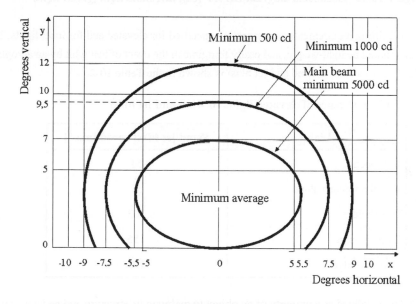

Figure 10-21 Isocandela diagram for runway edge light where width

of runway is 45 m (white light)

The requirements for the Thorn company's EL-AT threshold lights and their compliance are shown in Figure 10-22.

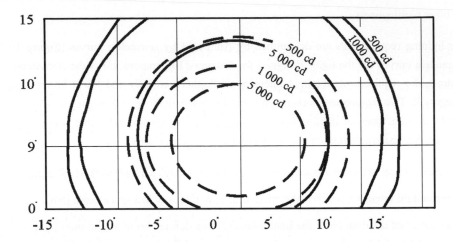

Figure 10-22 Isocandela diagram EL-AT [cd], threshold light (green light)

The mechanical fitting requirements are separately specified for elevated and for inset lights. The elevated lights should be light-weight, and easily frangible in the event of being hit by an aeroplane or a vehicle. The resistance to wind and jet blast is shown in the Table 10-2.

Table 10-2   The resistance of elevated lights to wind and jet blast

Light	Wind/jet blast speed [m.s^{-1}]
Approach	51
Visual approach slope	51
Threshold	103
Runway edge	155
Taxiway edge	103

The frangibility is defined as a property of an object to maintain its structure and resistance until a specified maximum force. If that force is exceeded, the mounting should deform or break in such a way as to create a minimum hazard for the aeroplane.

In approach systems, if required due to descending terrain, the so called safety masts are used. They are light-weight, usually lattice structures or towers made of aluminium or laminate, which will be frangible if hit by an aeroplane. At some airports, though the masts themselves are frangible, they are still mounted in substantial concrete plinths which can cause serious damage to an aeroplane.

The lights installed within approximately 2 m of the ground are supported by tubes or tripods attached to frangible fittings at ground level. The mechanical strength of the fitting should not be greater than 700 Nm for lights with the upper edge less than 360 mm over the level of the terrain. If the approach lights are higher than this, the strength of this light fixture may be correspondingly higher.

The inset lights should be resistant to dynamic effects caused by aircraft and ground vehicles passing over them. The basic mechanical requirement relates to the strength of the light cap expressed as a pressure applied on a surface unit. For a common inset light with a diameter of 300 mm, the required static strength of the cap shall be 230 kN. In addition, the temperature at the point of contact between the inset light and the tyre of an aeroplane created by temperature transfer, or by heat radiation of the lamp, should not exceed 160 °C during a 10 minute exposure.

A low-voltage cable from the insulation transformer to the light is designed for a voltage of 250 - 760 V. The insulation resistance of the light should be at least 2 MΩ. They must operate within a range of temperature from – 40 °C to + 55 °C. The fittings of elevated lights must be constructed so that the exchange of a damaged or broken light can be quick and uncomplicated.

With inset lights, the principal factors relate to the security of the fitting, the service life, the light source and speed of change of a cap with its optics *in situ*. The light bulb in the cap of an inset light should be changed in a workshop. It is also important that an inset light should resist damage by winter maintenance equipment like snow ploughs.

10.2.3 Lighting Systems Construction and Operation

10.2.3.1 Lighting Systems Design and Installation

Although ICAO and FAA have implemented a high degree of standardisation and uniformity into lighting systems in aerodromes, the installation of lighting systems is usually specific to each case. It is affected by the following factors:

✦ geometrical characteristics and arrangement of runways, taxiways and stands

✦ arrangement of power supply components (transformer substations) and the aerodrome control tower

✈ soil and climatic conditions

✈ previous habits and experience of the lighting systems operator, which are usually taken into account in the project

✈ anticipation of future additions to the system.

A renewal of the lighting systems is often connected with a reconstruction of the runway surface. The speed and quality of the works are influenced by the following factors:

✈ thorough preparation of the work, including the scheduling of work

✈ harmonisation of civil and electrical aspects of the work

✈ phasing the works to make the optimum use of climatic conditions.

Utilisation of an optimum season of the year is of principal importance, in particular in northern regions, where a favourable climate period is very short. Low temperatures and high air humidity have a negative impact in particular on installation of inset lights by gluing with epoxy resins into a concrete or asphalt pavement, and on insulation of cable couplings.

10.2.3.2 Maintenance of the Lighting Systems

The role of the lighting system maintenance personnel is to ensure that the equipment functions reliably of during its service life. This is effected by:

✈ the technical quality of the lighting systems

✈ the quality of the project design

✈ the quality of construction works

✈ the organisation and equipment for the lighting system maintenance

✈ the level of knowledge and skills of the maintenance personnel

✈ the timeliness and availability of information on the real condition of the equipment in operation.

The most important parameter of the lighting systems to be monitored is the number of serviceable lights. The requirements for ensuring the serviceability are specified by Annex 14. A light is considered serviceable whose light intensity in the required direction has not dropped under 50 % of the light intensity of a new one.

The requirement for the Category I systems is that at least 85 % lights are serviceable in:

+ precision approach system

+ threshold lights

+ runway edge lights

+ runway end lights.

Considerably more stringent requirements apply to the Category II and III systems. At least 95 % lights should be serviceable in the following sets:

+ inner 450 m of the Category II and III approach system– runway centre line lights

+ threshold lights

+ runway end lights

Furthermore, the following should be serviceable:

+ at least 90 % touchdown zone lights

+ at least 85 % lights of approach systems in a distance of 450 m from the runway threshold, and further

+ at least 75 % runway end lights.

Besides the above requirements, there are supplementary conditions, such as the requirement that no two consecutive failed lights shall occur in the runway and taxiway centre line systems. In the STOP- bar lights, only one light failure is allowed.

With the standard twenty year service life, the first cost of the lighting systems is a relatively small part of the total life-cycle costs. The quality of the equipment in its entirety is of prime importance.

The main effects that are directly connected with the costs on maintenance are:

+ the price and service life of the light bulbs

+ the resistance of the lights and fittings to corrosion

+ the resistance of the lights and fittings to the effects of air traffic, such as the impact of the jet engine combustion products, snow and ice cleaning equipment

+ the quality of the cables and connectors

+ laying the cables into cable ducts and the isolating transformers into easily accessible shafts

➜   the reliability of the constant current regulators and other electric equipment

➜   the reliability of the remote control and monitoring equipment, and the timeliness and availability of information on the lighting system condition

➜   low power requirements of the lights and the related costs of electrical energy.

These costs are all influenced by the costs of the materials and labour. It is best to use the same team of workers who also maintain other electric equipment. They should also participate in the installation of a new lighting system. In practice, the system of daily, weekly, monthly and half-yearly maintenance checks has been proved to be desirable.

The maintenance activities are recorded in service logs. The organisation of the maintenance is designed in the light of previous habits and climatic conditions. Where the climatic differences between summer and winter are big, and inset lights are in the majority, it is convenient to carry out extensive maintenance before the arrival of the winter season, including the complete replacement of lamps in the inset lights, and another extensive maintenance intervention in spring. The maintenance in winter months may be made easier by laying the cables into cable ducts, the isolating transformers into shafts, and by a proper construction of inset lights.

The effectiveness and speed of reaction in the maintenance of the lighting system is influenced also by equipment as well as personnel. It is necessary to have a suitable vehicle with radio communication and radio telephone. In the vehicle, equipment should be available for inspecting the cable network, a compressor and wrenches for maintaining the inset lights. The light intensity of inset lights is often impaired by impurities in the optical system, such as the remains of rubber from aircraft tires. In order to maintain a permanent light intensity to the required limits, it is necessary to clean the inset lights.Cleaning may be carried out by compressed air containing crushed walnut shells or olive stone pips, compressed liquids, mechanical brushes, or a combination of these. Full cleaning of the optical parts of inset lights is carried out in maintenance workshops when the light bulbs are replaced.

10.2.4 Trends in Lighting Systems Development

It may be expected that the future development will use the serial distribution of aerodrome lighting systems, except heliports, where a parallel supply system will be applied. The manufacturers will make an effort to produce unified equipment, including a wide variety of normative requirements. Reduction of power requirements of the lights, utilisation of light sources with long service life provides the manufacturers with a competitive advantage. Operationally undemanding inset lights of good quality will find their implementation in East European countries in particular, in connection with upgrading the aerodromes to higher categories of meteorological limits.

A modern lighting system at the beginning of the third millennium will be characterised by small power requirements, high service life of components and light sources, easy, quick and cheap maintenance. The control and monitoring systems will provide comfort for the operators and diagnostics of the condition for the needs of an effective maintenance operation.

11

ELECTRICAL ENERGY SUPPLY

Tony Kazda, Bob Caves and František Bělohradský

11.1 BACKGROUND

The consumption of electrical energy of a large international airport may be compared with that of a town with 15 000 to 30 000 inhabitants. Besides electrical energy required for the aids to air traffic operation such as lighting systems, radar, communication and meteorological systems, it is necessary to allow for electricity consumption in hangars, buildings and other airport facilities. This chapter will deal only with the supply and distribution systems of electrical energy for lighting systems and other equipment for air traffic operation.

11.2 ELECTRICAL SYSTEMS RELIABILITY

For lighting systems, telecommunication and radio-navigation aids, Annex 14 specifies the requirements for a standby electrical energy source in terms of the maximum switch–over time to a standby power supply. The requirement for a standby source may be fulfilled by constructing an independent supply of electrical energy to the aerodrome, connected with a 100 kV switching station, or by a standby source complying with the required parameters. The decision to construct an independent supply to the aerodrome should take account not only of investment costs, but also the electrical energy supplier's pricing policy and the use that may be made of surplus electrical energy.

On small aerodromes, the power to all the equipment may be covered by a central standby source. On large aerodromes, the number of installations that must be safeguarded increases, as do the requirements for power output of standby sources and the distances between the installations. The requirements for standby sources vary considerably in switching time and in the required power output of the standby source, according to the kind and category of the aerodrome. The greater the distance of the installation from the standby source, the lower is the reliability of the whole system as a consequence of possible failures in transmission lines. Therefore in larger aerodromes it is not practically possible to cover the entire consumption by a single, central standby source. For that reason, a decision on the number of standby sources and their arrangement should be made not only on the basis of an economic assessment, but consideration should be given also to the reliability of the electrical energy supply. For Category II precision approach runways, the probability of failure of electrical energy should not by higher than 1.10^{-1} per year. The standby sources, which are in the majority of cases diesel generators, are usually located directly in the transformer stations that feed the lighting systems. A special separate room is established for them, in which a constant temperature should be maintained, exhaust gases off-take and dispersion should be provided, and fuel store reliably designed to prevent fuel accident/incidents.

In the majority of cases, the radio-navigation and communication systems are powered by battery sources with automatic switching.

The requirements for a standby source of electrical energy and switching time are given in the Table 11-1.

In order to ensure those requirements are met, a whole range of technical solutions may be adopted. Besides an independent supply of electrical energy, some aerodromes have heating stations equipped with low pressure steam turbines and generators which can supply electrical energy to selected facilities. This is termed co-generation. At some airports, the heating/power stations serve as the main source of electrical energy and the connection to the 'public' high voltage line is the standby. The standby sources may be divided into the following groups:

→   rotational standby sources - diesel generators:

→   with manual start – up within 2 minutes

→   with automatic start -within 15 seconds

→   short-break -within 1 second

→   no-break -uninterrupted power supply.

Table 11-1   Secondary power supply requirements (*Source: Annex 14, Aerodromes*)

Runway	Lighting aids requiring power	Maximum switch-over time
Non-instrument	Visual approach slope indicators	2 minutes
	Runway edge	2 minutes
	Runway threshold	2 minutes
	Runway end	2 minutes
	Obstacle	2 minutes
Non-precision approach	Approach lighting system	15 seconds
	Visual approach slope indicators	15 seconds
	Runway edge	15 seconds
	Runway threshold	15 seconds
	Runway end	15 seconds
	Obstacle	15 seconds
Precision approach Category I	Approach lighting system	15 seconds
	Runway edge	15 seconds
	Visual approach slope indicators	15 seconds
	Runway threshold	15 seconds
	Runway end	15 seconds
	Essential taxiway	15 seconds
	Obstacle	15 seconds
Precision approach Category II/III	Approach lighting system	15 seconds
	Supplementary app. lighting barrettes	1 second
	Obstacle	15 seconds
	Runway edge	15 seconds
	Runway threshold	1 second
	Runway end	1 second
	Runway centre line	1 second
	Runway touchdown zone	1 second
	All stop bars	1 second
	Essential taxiway	15 seconds
Take-off RWY for RVR < 800 m	Runway edge	15 seconds
	Runway end	1 second
	Runway centre line	1 second
	All stop bars	1 second
	Essential taxiway	15 seconds
	Obstacle	15 seconds

Static standby sources:

✈ battery short-break -within 1 second

✈ battery no-break -uninterrupted power supply.

Diesel generators with manual start–up within 2 minutes are used in smaller aerodromes with a non-instrument or an instrument runway. On larger aerodromes, they may be used for the provision of supplies of electrical energy to individual facilities such as the lighting of buildings, lifts, and the like. In general, they provide outputs up to 2.5 MVA. In emergency cases, the diesel generators with manual start may also be used, as may all the remaining standby sources (short/no-break), for ensuring operation of the lighting systems under Category II or III conditions. When the meteorological limits of Category II are reached, the generator is started, and is in operation during the whole time that the corresponding conditions last. In the event of a failure of the network, an immediate switch over to a standby source is made. When fuel prices are high, such a solution should only be chosen if there is no alternative.

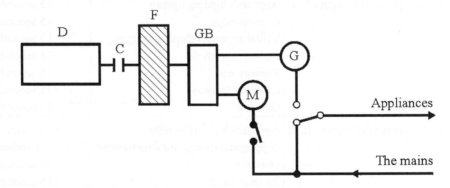

Figure 11-1   Short-break diesel generator
D – diesel generator, C – clutch, F – flywheel, GB – gearbox, M – motor, G - alternator

Diesel generators with automatic start within 15 seconds are connected with an automatic control box which monitors the parameters of the electrical network. When the specified voltage and frequency parameters of the network drop under the determined limit, the automatics will start the diesel generator and switch the supply over to the standby source. If the first start is unsuccessful, the automatics will repeat the start procedure after several seconds. Diesel generators with automatic start are used not only as separate equipment, but also in combination with battery sources (see below), where a battery provides an immediate back-up of electrical energy in the event of a failure of the network, and after starting, the diesel generator will charge the battery.

A short-break diesel generator (Figure 11-1) has a heavy flywheel, which is driven at a constant speed through a gearbox by a electric motor supplied from the mains. An alternator, which in the normal state of the network runs with no load, is on one side of the flywheel shaft. On the other side of the shaft, there is a diesel generator connected by means of an electromagnetic clutch. When the monitored parameters of the electric network drop under the determined values, the diesel generator is automatically started and connected by means of the electromagnetic clutch to the flywheel and alternator. While the diesel generator reaches its full power, the alternator is driven by the turning flywheel. At the same time the automatics switch the appliances over to the alternator terminals. In the event of a failure of electrical energy, there is a short-term voltage drop of about 10 %, and a frequency drop of about 4 %.

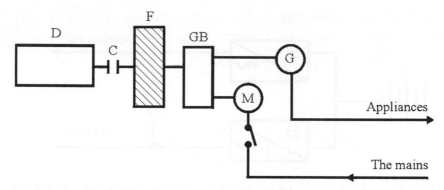

Figure 11-2   Not-break diesel generator
D – diesel generator, C – clutch, F – flywheel, GB – gearbox, M – motor, G - alternator

The no-break diesel generator (Figure 11-2) is used when no short-time electric current failure is permitted, for example in a supply of current for computer systems or secondary radar. The no-break equipment differs from the previous type in that the electric motor constantly drives not only the flywheel, but also the alternator, from which electricity is supplied to appliances. In the event of an interruption of the mains supply, there will be only a very short-term voltage and frequency drop.

The short-break and no-break equipment is expensive to purchase and to operate. This is not only due to requirements to provide a high reliability of the system, but also due to the need for high power inputs for the lighting systems of higher categories of landing aids. Some firms supply equipment that may be used differently according to meteorological conditions, normally as a diesel generator with automatic start, and when meteorological conditions drop under the determined limits, it may be switched over to short-break and no-break modes.

In static battery standby systems, the basic source of electrical energy is an accumulator battery with the necessary capacity. The endurance required before the diesel generator starts may vary from several minutes up to several hours. Modern batteries with a closed cycle have a high service life up to 20 years, and high capacity with small dimensions. They are cheap and undemanding to maintain. It is possible to centralise supply for the systems with a fall-out not longer than 1 second to a single transformer station, and to back it up with a standby battery within 1 second. In spite of the fact that the static standby sources are used mostly for appliances with low consumption, such as radio stations, radio-navigation equipment, etc., they may also provide very high outputs for a short time. In addition, as mentioned in Chapter 10, modern lights have ever lower and lower power requirements. A static standby source may be designed as short-break or no-break according to how it is connected (see Figure 11-3).

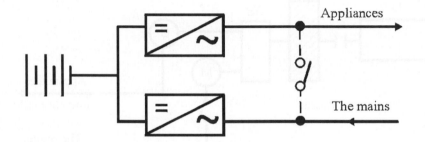

Figure 11-3   Static standby source; no-break – full line, short-break – dashed line

11.3 SUPPLY SYSTEMS

The supply of lighting systems may be provided in two ways, namely by parallel or serial distribution systems.

11.3.1 Parallel System

The well-established parallel distribution system (Figure 11-4) is used at present for small lighting systems. It uses a constant voltage source. A three-core cable is connected to the switchboard of the transformer station equipped with a stepping control transformer for reducing voltage. Every individual light is connected to cable conductors by means of the isolating transformers. An advantage of the parallel distribution system is its simplicity. The maintenance personnel need not be qualified for work with high voltage. The current in halogen lamps of the parallel distribution system is not regulated. As the tungsten gradually settles on the lamp filament, the lamp resistance drops and the current in the lamp increases, the tungsten melts, evaporates, and the lamp acquires its specified parameters. These lamps give a constant light during their whole service life. The main disadvantage of the parallel distribution system is that the voltage drops with distance from the transformer station, and thereby also the intensity of the more distant lights drop. For that reason, the parallel distribution system is only suitable for small areas, such as the lighting systems of heliports or aprons.

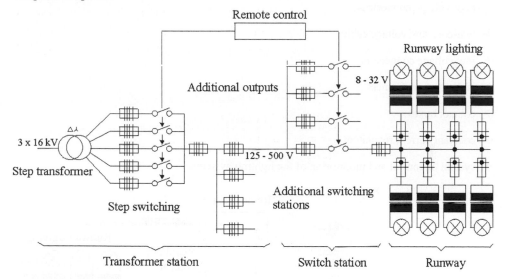

Figure 11-4   Parallel distribution system - principle

11.3.2 Serial System

11.3.2.1 Serial System – the Principle

The serial distribution system (Figure 11-5) is now almost universally adopted for runway lighting systems. Contrary to the parallel distribution system, which is supplied with constant voltage, the serial distribution system operates with constant current. It consists of a closed loop of single-core cable. Isolating transformers are put into the circuit, which ensure that there will not be a break of the primary circuit when the secondary circuit is interrupted (lamp bulb failure) and that the remaining lights continue to operate.

11.3.2.2 Serial System – Components

The distribution cable, the quality and reliability of particular components, and the quality of installation all contribute to the reliability of the lighting systems. The components of a serial distribution system are:

→ single-core high voltage supply cables

→ high voltage connectors

→ two-core low voltage cable

→ low voltage connectors

→ isolation transformers

→ contact boxes

→ constant current regulators

→ system of control and monitoring of the lighting systems.

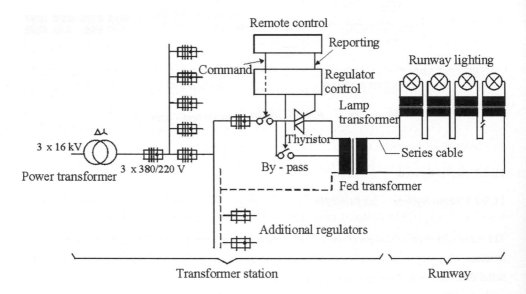

Figure 11-5   Serial distribution system - principle

The number of elements in a lighting system increases with its category. Therefore for lighting systems of higher categories, elements of better quality should be used in order to avoid a loss of reliability of the whole system. The system of a Category II runway includes up to 100 km of single-core high voltage cables, approximately 1 000 isolation transformers, 2 000 high voltage connectors, and a large number of low voltage connectors.

High voltage cables are normally single-core cables with a copper core with a cross section of 6 mm. They are designed for an operating voltage from 500 to 5 000 V. The usual range of operating voltages is 500, 600, 750, 1200, 3000, 5000 V. The cables are unshielded at the lower voltages. Shielded cables limit the interference caused by the thyristor regulation of the constant current regulators. Cables can be made in exact lengths, and equipped with standard connectors while in the production plant. Cable is more often supplied on drums, and connectors are mounted on site at the aerodrome. If the connectors are made in the production plant, the final mounting of the lighting system can be completed more quickly.

Advantages of a cable system with connectors mounted in the aerodrome are as follows:

✈ simplification of design works

✈ simpler production

✈ saving of up to 10 % of cable length

✈ easier installation when cable ducts and transformer shafts are used, since the required lengths may be provided with a centimetre precision.

Current is fed from the isolation transformer to the lights via low voltage (LV) cables. The LV cables are equipped with standard FAA connectors. In the event of the light failing, the connector is severed, and the light disconnected from the cable. If the light is distant from the isolation transformer, an extending LV cable (flexo) is used. This is a two-core LV cable equipped with a standard termination and FAA socket. LV cables are designed for a voltage of 250 or 600 V, while the operating voltages are usually between 6 and 45 V.

The isolation transformers are enclosed in rubber or plastic material, which ensures that they remain water-tight. The isolation transformer is provided with two HV leads for connecting two power cables and one LV lead for connecting to a light. The transformer output range is 30, 45, 65, 100, 150, 200 a 300 VA. The HV leads are designed for a voltage of 5000 V, the LV lead for a voltage of 250 or 600 V.

The lighting systems are supplied from constant current regulators. These are power supply transformers with adjustment and monitoring circuits. The adjustment of the intensity of the lights is most often in 5 steps that are 1 %, 3 %, 10 %, 30 %, 100 % of light intensity. In some systems, such as PAPI and taxiway side lights, the light intensity is adjusted in three steps of 10 %, 30 % a 100 %. The 7-step adjustment, with 80 % light intensity, is used only occasionally. That setting ensures high light intensity, and does not reduce the service life of the light sources. The monitoring system provides information about the condition of the constant current regulators in terms of condition of the loop's insulation and the number of failed lights to indicate the need for maintenance of the lighting systems and to alert the air traffic controllers. The constant current regulators are placed in transformer stations and supplied from LV switchboards. Connector boxes, which allow several circuits of the lighting systems to be connected to a single constant current regulator, are also part of the supply system.

The constant current regulator maintains constant current of a specific value throughout the whole circuit despite a varying load, caused, for example, by burnt-out filaments in the lamps. The constant current regulators usually have an output range of 4, 8, 16, and 32 kVA. In more advanced lighting systems, which have a low power demand, the power output range is usually 2.5, 7, 7.5, 10, 15, 20 and 25 kVA, though power outputs over 15 kVA are rarely used. In the power input loop of 15 kVA, a power cable with an operating voltage not higher than 3000 V may be used. Output current of constant current regulators is usually either 6.6 or, more rarely, 8.3 A at the maximum intensity of the lights. At lower degrees of light intensity, the output current is proportionally lower.

A constant current regulator must indicate certain conditions:

✈ operation

✈ failure

✈ insulation condition of the loop

✈ number of failed lamps

✈ degree of light intensity in the loop.

A clock records the number of hours the loop operates at a maximum light intensity and the total number of hours of operation. In simple systems, the constant current regulators usually indicate only operation, failure, and insulation condition of the loop.

The constant current regulators are equipped with over-current and over-voltage protection, which automatically disconnect them so that the lights cannot burn out if the power supply cable or separation transformers are damaged. An electric charge may be induced in the long loops of the lighting systems when a lightning strikes near the lights during a storm, which may damage the constant current regulators. Therefore the loops are connected with the constant current regulators by means of a lightning arrester. Another possible solution is installation of a strong non-insulated and grounded conductor in the ground over the power cables.

In some cases it is convenient to supply several circuits of the lighting systems from a single constant current regulator, for example two PAPI systems, STOP lighting bars, various sections of taxiway inset lights. Connector boxes placed beyond the constant current regulators, or in modern lighting systems, directly built into the constant current regulators, serve for switching of the specific loops. The boxes are equipped with HV connectors. On the basis of instructions from the control tower, the particular circuits are connected according to the requirements and pre-determined algorithms. For example, the runway centre line inset lights are connected to two circuits. Only one of them is always switched on, according to the runway that is in use. The runway centre line inset lights are lit only in the direction either of landing or of take-off. The PAPI systems are similarly supplied from constant current regulators for both directions of landing. In the case of taxiways, the connector boxes operate in a different way. According to the instructions from the control tower, one, two and even more circuits whose lights may be lit simultaneously are connected. A typical example may be supplying two STOP-bar lights from a single constant current regulator.

Modern constant current regulators are controlled by a microprocessor. The microprocessor regulates and monitors all the conditions and protections of a regulator. A more reliable and simpler communication system is also provided between a constant current regulator and the control and monitoring system, including an unambiguous identification of a failure of the regulator indicated in the office of the maintenance manager.

In order that the reliability may be improved, the lighting system supply is provided by a whole range of circuits. Depending on the number of insulation transformers in the loop, their power input, and HV cable length, the power input of the constant current regulator may be determined. The biggest load is usually in the loops of an approach system of high light intensity, and in the loops of runway edge lights. Every such loop supplies 60 to 100 lights with a power input of 150 to 200 W, and the loop power input may reach up to 20 kVA. Small lighting systems, e.g. STOP-bar lights, PAPI and simple approach lighting systems require relatively small power input. The power input of the lights, including losses in the separation transformers, is given in Table 11-2.

The losses in the cable reach approximately 200 W per 1 km of the cable length. The calculation of the power input of the loop of an approach system of high light intensity are given in the following example.

Example – power input of the loop:

$$P = A \times B + L \times Z = 65 \times 177 + 3{,}2 \times 200 = 12\ 145\ \text{VA}$$

where:

P – power input of the loop

A – number of lights   (65 pcs)

B – power input of the light   (177 VA)

L – cable length   (3.2 km)

Z – loss in the cable   (200 W.km^{-1})

In this case, the calculation shows that the load from the lights is decisive, while the losses in the cable are of little importance.

Table 11-2   Power inputs of the lights

Light bulb [W]	Power input [VA]
45	56
65	77
100	118
150	177
200	230
300	354

11.4 ELECTRICAL SUPPLY TO CATEGORY I - III LIGHTING SYSTEMS

The lighting system of an aerodrome contains a whole range of systems which should be reliably supplied and controlled. Each aerodrome has its specific features. The electrical supply for a lighting system of low or middle light intensity of a non-categorised aerodrome is usually simpler, and contains from 2 to 8 constant current regulators. On the other hand, Category II and III systems are provided with a more extensive supply system containing a greater number of constant current regulators. For a Category II aerodrome, the usual number of constant current regulators is 15 to 25 in one transformer station. The electrical supply to the extensive taxiway systems in Category III is very complicated. The centre line guidance in a runway with inset lights is divided into a whole range of separately supplied sections, that is a large number of constant current regulators and connector boxes. For example, Munich airport, equipped with the lighting systems for Category III, is divided into 1200 separately supplied circuits.

12

AIRPORT WINTER OPERATION

Tony Kazda and Bob Caves

12.1 SNOW AND AIRCRAFT OPERATION

Snow means pleasure for children but it is also the source of problems for airport administrators. The obligation to maintain a clear surface of the runway and other movement surfaces of the airport has been imposed upon the airport operator by Annex 14. Contamination of the movement areas, which include e.g. snow and ice, mean the limitation or closing of an airport's operation. Icing on the aircraft and its dispersal may decrease utilisation and disruption of the flight schedule. This results in loss of revenue for the airports and airlines and also in costs connected with clearing the snow and ice. Each airport has to:

→ provide an effective snow plan

→ provide regularity of flight operation in the winter despite adverse meteorological conditions.

When clearing ice by chemical means, the negative impact of chemical substances on the environment must be minimised.

A layer of snow on the runway surface causes:

→ resistance acting on the aircraft's wheels during the take-off run; the magnitude depending on the density and thickness of the snow layer, characteristics of the aircraft undercarriage, and the speed and weight of the aircraft

✈ increase of drag and decrease of lift of the aircraft during the take-off run due to snow being thrown up from the wheels, particularly the nose wheel

✈ decrease of braking effect from the runway surface friction, particularly when the contamination is ice, increases the possibility of exceeding the available distances for landing or rejected takeoff.

Icing on the aircraft, particularly on the lifting surfaces, changes the aerodynamic characteristics and flight performance, and it can block or impair flight controls and increase the weight of the aircraft. Iced sensors can cause the pilot to receive wrong information about speed or engine condition. Therefore before take-off all the ice has to be removed.

The effect of contamination on the movement areas of the airport on the aircraft operation depends on several factors, in particular:

✈ air temperature

✈ runway temperature

✈ specific density of snow.

The higher the air temperature the higher is also the specific weight of the snow. Three kinds of snow have been defined according to its specific weight:

Dry snow. Snow which can be blown if loose or, if compacted by hand, will fall apart again upon release. Specific weight of dry snow is up to but not including 350 kg.m^{-3}.

Wet snow. Snow which, if compacted by hand, will stick together and tend to form a snowball. Specific weight from 350 kg.m^{-3} up to but not including 500 kg.m^{-3}.

Compacted snow. Snow which has been compressed into a solid mass resisting further compression and which will hold together or break up into lumps if picked up. Specific weight of compacted snow is above 500 kg.m^{-3}.

Slush is snow saturated by water. The snow is so saturated by water that the slush splashes around when it is stamped upon. The specific weight of slush is from 500 kg.m^{-3} up to 800 kg.m^{-3}.

The easiest way to ascertain the specific weight is by taking a sample of a certain volume of snow and dissolving it. The higher the specific weight of snow or slush the higher is the resistance acting on the aircraft's wheels. A layer of slush on the runway of approximately 4 mm is enough to affect the performance of an aircraft, and if the slush layer is above 13 mm the runway has to be closed and cleared. For comparison the runway must be closed if a layer of dry snow is 5 cm thick.

12.2 SNOW PLAN

The snow, slush or ice must be removed from the movement areas of the airport quickly and without residues in order to ensure safe operation of aircraft. Economic factors must also be considered. The snow plan of an airport should specify the organisation, the provision of airport equipment and the necessary chemicals, and co-ordination of work with air traffic control. The possible scope of a winter plan is given in Airport Services Manual, Part 2 Pavement Surface Conditions, Chapter 7 Snow removal and Ice Control (ICAO Doc 9137-AN/898).

Preparation of the whole airport for the winter is very important. It includes not only full preparation of equipment, training of new workers and retraining of full time personnel, but also the maintenance of the movement areas. The basic requirement is that the airport and all equipment must be in perfect condition sufficiently in advance of the possible appearance of adverse weather conditions. The training of workers, who will perform the winter maintenance should include:

✈ Radiotelephonic procedures. The workers must have necessary qualifications for operation of the radio station, they must be familiar with the operation of the given type of transmitter and know the radio phraseology.

✈ Procedures for removing of snow and ice. The procedures are different for each kind of snow and the runways in use.

✈ Operation of equipment. Each worker must control properly the assigned technical equipment in any weather by day or night without affecting safety of operation.

✈ Airport. The workers must know perfectly all parts of the movement areas of the airport.

Many workers from various departments of the airport participate in the winter maintenance and it is necessary to co-ordinate their work. The snow co-ordination committee controlling individual activities of winter maintenance will include airport management, meteorological services, air traffic controllers and representatives of airlines.

When preparing the snow plan it is necessary to consider a number of factors such as topography, climatic conditions, airport location, types of aircraft that use the airport, density of operations and physical characteristics of airport movement areas.

Priorities for clearing snow and ice from various parts of movement areas have been specified in

Annex 14 and they must also be entered in the winter plan. They can be changed, but only after an agreement between the airport operator and air traffic control. Priorities for clearing movement areas are as follows:

→ runway(s) in use

→ taxiways serving the runway(s) in use

→ apron

→ holding bays

→ other areas.

In addition to the above areas, attention has to be paid to clearing snow from the vicinity of antennae, radio navigation equipment, and particularly from the vicinity of the glide path ILS antenna, the signal of which is very sensitive and could be distorted by the layer of snow. Depending on weather conditions and operational possibilities, other movement areas and access roads will also gradually be cleared as the opportunity arises.

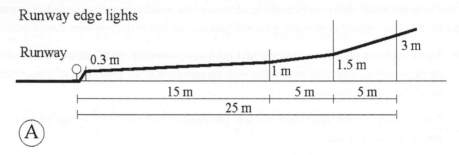

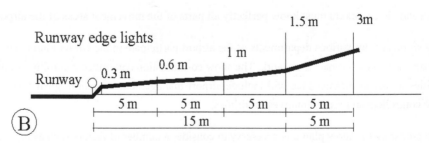

Figure 12-1   Maximum height of snowbanks
A – Runways used by very large aircraft (such as B-747, DC-10, L-1011)
B – Runways used by other than very large aircraft

Because snow is hygroscopic, its weight can grow quickly when the temperature is around zero. When clearing of snow is being planned, it is necessary to take account of this property. Therefore with the fall of the first snow flakes, clearing of snow should already be started on the most important areas. Usually it is still possible to allow a few more aircraft movements before the runway will have to be closed and the centre of the runway be cleared. The procedure used for the first clearing of the runway depends on the equipment which is available, the kind of snow (dry, wet, compacted or slush) speed and direction of the wind and other factors. Which of the runways shall be cleared will usually be specified by the airport duty manager in co-operation with air traffic control on the basis of a meteorological forecast. If critical snowbanks are not formed (see Figure 12-1) it will be possible to open the airport for operations within a relatively short time. If snow continues to fall, further closure and clearing of the runway become necessary. It is no use clearing the runway during a snowstorm; the snow will be blown onto the runway as fast as it is being removed. After the snowstorm is over, and the airport has to be opened as quickly as possible, the personnel are tired and the equipment needs replenishing and inspection. If two runways are available and if this is made possible by the meteorological conditions, it is advantageous to open one runway and continue to clear snow from the other.

It is not economically realistic to maintain a sufficient quantity of equipment to cope with the very worst predicted weather. Therefore, if those extreme conditions occur, the airport will have to be closed.

The aircraft operators are informed bout the up-to-date condition of the airport or its closure by means of SNOWTAMs, which are distributed by means of the AFTN telex network.

If the airport has only one runway, it is often necessary to use equipment which enables the snow to cleared at high speed. For other areas with lower priority it is possible to use normal procedures of clearing snow and ice or less used areas can be temporarily closed.

Although keeping open the aircraft movement areas has the highest priority, it is necessary to ensure that the airside and landside roads and parking places are available in order to ensure transportation to and from the airport. It is sensible to contract-out the clearance of such areas e.g. by agricultural companies or by construction companies, since their equipment is little used in the winter.

In order to remove snow and ice from the movement areas it is possible to use mechanical, chemical or thermal means.

SNOWTAM	Priority GG Indicator		ADDRESSES	EDZZSA LIZZSA ABZZSC		
Date and Time of filing 01 02 0930		Originator's Indicator EALLYN		SNOW NOTAM ("S" SERIES) SERIAL NUMBER 279		NOTAM S

AERODROME	A	EACD				
DATE/TIME OF OBSERVATION *(Time of completion of measurement in GMT)*	B	01020850	B	01020905	B	
RUNWAY DESIGNATORS	C	16/34	C	07/25 .	C	
CLEARED RUNWAY LENGTH, IF LESS THAN PUBLISHED LENGTH *(m)*	D	3300	D	----	D	
CLEARED RUNWAY WIDTH, IF LESS THAN PUBLISHED WIDTH *(m; if offset left or right of centre line add "L" or "R")*	E	40L	E	----	E	
DEPOSITS OVER TOTAL RUNWAY LENGTH *(Observed on each third of the runway, starting from threshold having the lower runway designation number)* NIL – CLEAR AND DRY 1 – DAMP 2 – WET or water patches 3 – RIME OR FROST COVERED *(depth normally less than 1 mm)* 4 – DRY SNOW 5 – WET SNOW 6 – SLUSH 7 – ICE 8 – COMPACTED OR ROLLED SNOW 9 – FROZEN RUTS OR RIDGES	F	4/5/4	F	57/56/57	F	
MEAN DEPTH *(mm)* FOR EACH THIRD OF TOTAL RUNWAY LENGTH	G	20/10/20	G	05/05/05	G	

BRAKING ACTION ON EACH THIRD OF RUNWAY AND MEASURING EQUIPMENT
  MEASURED OR CALCULATED COEFFICIENT *or* ESTIMATED BRAKING ACTION
  0.40 and above GOOD – 5
  0.39 to 0.36 MEDIUM/GOOD – 4
  0.35 to 0.30 MEDIUM – 3
  0.29 to 0.26 MEDIUM/POOR – 2
  0.25 and below POOR – 1
  9 – unreliable UNRELIABLE – 9
 (When quoting a measured coefficient use the observed two figures, followed by the abbreviation of the measuring equipment used. When quoting an estimate use single digits)
	H	30/35/30 MUM	H	32/36/9 MUM	H	

CRITICAL SNOWBANKS *(If present, insert height (cm)/distance from the edge of runway (m) followed by "L", "R" or "LR" if applicable)*	J	30/5 L	J	-------	J	
RUNWAY LIGHTS *(If obscured, insert "YES" followed by "L", "R" or both "LR" if applicable)*	K	YES L	K	-------	K	
FURTHER CLEARANCE *(If planned insert length (m)/width (m) to be cleared or if to full dimensions, insert "TOTAL")*	L	TOTAL	L	-------	L	
FURTHER CLEARANCE EXPECTED TO BE COMPLETED BY . . . *(GMT)*	M	1300	M	------	M	
TAXIWAY *(If no appropriate taxiway is available, insert "NO")*	N	NO	N	-------	N	
TAXIWAY SNOWBANKS *(If more than 60 cm, insert "YES" followed by distance apart, m)*	P	YES 10	P	-------	P	
APRON *(If unusable insert "NO")*	R					
NEXT PLANNED OBSERVATION/MEASUREMENT IS FOR . . . *(day/month/hour in GMT)*	S	01021400				
PLAIN LANGUAGE REMARKS *(Including contaminant coverage and other operationally significant information, e.g. sanding, deicing)* RWY CONTAMINATION 100 % BOTH RWYS	T	RWY 07 SANDED, LAST 300 M RWY 16 COVERED BY 50 MM SNOW				

SIGNATURE OF ORIGINATOR *(not for transmission)*

Figure 12-2   SNOWTAM format

Source: Airport Service Manual, Part 2 Pavement Surface conditions

12.3 MECHANICAL EQUIPMENT FOR SNOW REMOVAL AND ICE CONTROL

Mechanical equipment for snow and ice clearing from the runway surface should always, whenever possible, be preferred. Its primary advantages, in comparison to chemical and thermal means, are lower costs and negligible environmental impact. Mechanical equipment is used mostly for snow clearing. There is little that can be done mechanically to treat a layer of ice but, under certain conditions, it is possible to improve the braking action by sanding the ice layer.

The speed and quality of snow clearing from movement areas of the airport depends to a great extent on the number and capacity of the equipment. When selecting the equipment the airport operator must consider a number of factors:

✈ scope of operation

✈ capital and operating cost of the equipment

✈ dimensions of airport areas

✈ availability of spare parts and possibility of repairs

✈ climatic conditions.

At small airports with general aviation operation or with only a few scheduled traffic movements a day, it may be best to outsource snow clearing. Alternatively, if it snows only occasionally, it is possible to close the airport for a few hours or days.

Airport manoeuvring areas have different physical characteristics from roads, so they require special equipment, which enables faster and higher quality clearing of the runway surfaces. The airport area is flat and the runways are wide. In the runway there are usually inset lights protruding above the surface of the runway. The terrain behind the runway edge lights, which is often not paved, must also be cleared. The snow layer on the airport runway is usually not deep, but the time for clearing the runway is substantially shorter than for roads. All these are characteristics which have to be considered when selecting the mechanical snow removal equipment.

The number of pieces of equipment for winter service can be specified on the basis of the average height of snow cover, amount of snow falling within one snowfall, area of the movement surfaces which are to be cleared, and the type and intensity of the air operations. It is usually not worthwhile to equip airports having a total snowfall below 40 cm a year and up to 5 cm of snow in one snowfall with a great number of expensive and high performance pieces of winter service equipment and vehicles.

Frankfurt airport has the fleet of 28 sweeper/blowers, 9 plough/sweeper/spreaders, 14 deicing vehicles and 3 sand spreaders. There are 720 000 m^2 of runway, 972 000 m^2 of taxiway, 4 130 000 m^2 of apron, 479 000 m^2 roads on airport and 231 000 m^2 of public roads. Snow falls on 16 days, average depth is 4 cm, max .in 24 hrs is 2 times 10 cm, 31 days of icing only (source The ACI 1999 Winter Services handbook).

The snow accumulated at the edge of the runway must be removed. Snow blowers are normally used for this. The performance of the snow blower is a critical parameter in the fleet of winter service equipment and vehicles, and it is also the most expensive equipment. At some airports in temperate climates and only a small layer of snow cover a year, it is sometimes possible to use a special type of plough which removes snow at high speed, rather than a snow blower. The use of such a plough has to be considered separately for each case, because several other factors have to be taken into account.

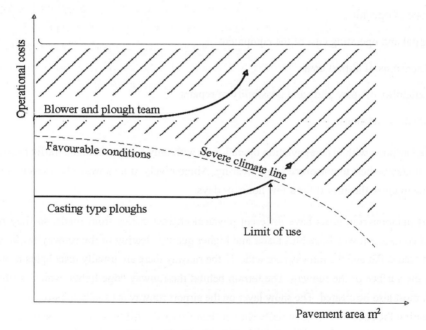

Figure 12-3   Casting type ploughs
Source: Airport Services Manual, Part 2 Pavement Surface conditions

Other winter service vehicles and equipment which are used on larger airports are air blower machines, ploughs, sand/aggregate trucks, chemical spreaders, tankers and loaders.

When removing small layers of snow it is best to use air blower machines to sweep the snow, since they will clean the surface of the runway properly. The height of the rotary brush above the surface of the runway is electronically controlled so as to clean the runway properly without the brush being excessively worn.

There are many types of special ploughs for airport use. The widest are wider than any of the road ploughs, at 4.5 m and more. The blade portion of some airport ploughs has been divided into several sections, each of them being separately spring loaded. If it contacts an obstacle, e.g. an inset light, the appropriate section will lift. Modern blades are produced of light composite carbon materials, which have three main advantages. They are substantially lighter than the common blades made of steel, their friction coefficient with the surface of the runway is low and they do not corrode. The low weight of the blade and low friction coefficient contribute to decrease fuel consumption.

An airport with regular air transportation should be equipped with one or more high performance snow blowers, which are capable of throwing snow with a the specific weight of 400 kg.m^{-3} a distance of at least 30 m. On airports with regular commercial transport, there should be sufficient equipment to ensure removal of a 2.5 cm snowfall from the main runway and from two most used taxiways connecting the main runway with the apron. On the basis of operational experience, clearing the taxiways requires approximately 25 % more time than clearing a runway of comparable area.

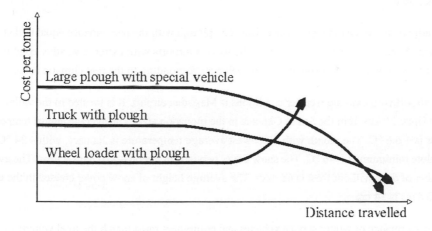

Figure 12-4 Ploughing costs/distance
Source: Airport Services Manual, Part 2 Pavement Surface conditions

The 2.5 cm thick layer of snow should be removed from the main runway and from the two taxiways within the following time limits with the stated number of air transport movements (atm):

✈ 40 000 and more atm, the snow should be removed within 30 minutes

✈ 10 000 to 40 000 atm, the snow should be removed within one hour

✈ 6 000 to 10 000 atm, the snow should be removed within two hours

✈ 6 000 and less atm, the snow should be removed within two hours, where possible.

Two ploughs should be included into the work group in front of each snow blower, the performance of which should correspond to the performance of the snow blower. It is necessary to close one runway system for only 20 minutes at Munich.

The equipment for a general aviation airport, which mostly serves aircraft up to 5 700 kg, vehicles and equipment for winter operations should ensure the removal of 2.5 cm snow layer from one runway, taxiway connecting the runway with the apron and from 20 % of the apron. The airport should be equipped by one snow blower capable of throwing snow of a specific weight 400 kg.m^{-3} at least 15 m. The snow layer 2.5 cm thick must be removed as follows:

✈ 40 000 and more movements a year, the snow should be removed within two hours

✈ 6 000 to 40 000 movements a year the snow should be removed within four hours

✈ 6 000 and less movements a year the snow should be removed within four hours, where it is possible.

The airport should be equipped with at least one plough with the performance equal to that of the blower. A higher value than 2.5 cm must be used at airports with extreme weather conditions as the critical height of snow conditions, the value corresponding to the real climatic conditions.

One airport with extreme weather conditions is Magadan airport. It is located in the valley of the river Uptar 30 km from the Sea of Okhotsk in the monsoon area. The average annual temperature of air is + 6.6 °C. The month with the lowest average temperature is January, with – 24 °C. The absolute minimum is – 46 °C. The snow cover persists on average 210 days a year. The average number of days without frost is 68 days. The average height of snow cover created in the course of 10 days is 54 cm.

The performance of winter service vehicles and equipment must match the total volume of snow, that needs to be removed from the movement areas. Because each airport has a set time for snow removal according to its position in the appropriate group and the reference height of snow cover of 2.5 cm has also been fixed, the area of the movement areas from which the snow is to be removed is the determinant of the capacity of winter service vehicles and equipment. The relationship is linear, as shown in Figure 12-5. The required performance of other vehicles and equipment can be determined in a similar way to the performance of the snow blower.

Table 12-1   Determination of snow blower capacity

1. Pavement area to be cleared		
Main runway	3 000 x 45 m	135 000 m²
Parallel taxiway with connectors	3 000 x 23 m	69 000 m²
Two connector taxiway	400 x 23 m	9 200 m²
Holding areas	2 x 40 x 60 m	4 800 m²
Apron	25 % of 16 000 m²	4 000 m²
Total area to be cleared		222 000 m²

2. Other parameters	
Snow removal time (1st group of airports)	30 min.
Temperature	- 4 ° C
Snow density	400 kg.m^{-3}
Depth of snow	2.5 cm

3. Determination of snow blower capacity		
Snow volume	222 00 x 0.025	5 550 m³
Snow mass	5 500 x 400	2 200 000 kg
Snow blower capacity per hr		4 400 000 kg
Snow blower capacity per hr in tonnes		4 400 t

The performance of the snow blower must be at least 4 400 t of snow per hour. The performance data of snow blowers given by the producers must be considered as approximate and in most cases they are optimistic. The performance must be verified by a practical test under the real conditions in the airport unless references can be obtained from other airport operators. In practice, the capacity of the snow blower can be 40 to 50 % lower than the given capacity.

The performance of other equipment must be higher that the performance of the snow blower. Their selection is affected by the same factors as the selection of the snow blower.

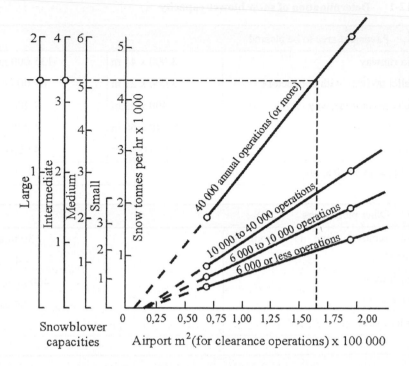

Figure 12-5   Snow blower selection card
Source: Airport Services Manual, Part 2 Pavement Surface conditions

In some countries the use of sand is permitted to improve braking action on the runway. However, a small grain of sand can damage an engine blade if sucked into it, so sand should be used as the last resort when it is not possible to use other chemical or mechanical means to remove the ice. Therefore sand is used primarily at very low temperatures, when the use of chemicals is not effective. The kind of sand must be carefully checked. Fine sand which can fall through a sieve with 0.297 mm mesh does not guarantee the required improvement in braking, and it can be easily blown away by the wind from the surface of the runway. Coarse material, which does not get through a sieve with 4.76 mm mesh can damage the jet engine blades and it also damages the leading edges of propeller blades. Individual sand particles must be sufficiently hard to resist the shear forces upon braking, but soft enough to minimise the damage of the metal surface of the aircraft structure. The sand must not contain small rocks, impurities, salts and other corrosive substances. The most suitable material is crushed limestone with knife-edges.

In order to increase braking action significantly and at the same time to minimise the possibility of sucking the sand into the engine, individual grains of sand must be embedded into the surface of the runway. This is possible only if the temperatures are well below zero. The possible manners of embedding the sand are:

✈ heating the sand before distributing it

✈ melting the ice sprinkled with sand by burners

✈ spraying water on ice sprinkled with sand.

After the grains of sand have been embedded into the ice, for which several tens of minutes are necessary, free grains of sand must be swept away and vacuum cleaned. The runway has to be cleaned in the same way after a thaw.

12.4 CHEMICALS FOR RUNWAY DE-ICING

Chemicals used for removing ice or for the prevention of icing on the surface of the runway must meet a number of requirements, which are often conflicting. They must be cheap and effective. They must damage neither the aircraft structure nor the runway. They must not be toxic and their harmful effects on the environment must be minimal. After de-icing chemicals are applied, a thin layer of water is formed over the remaining sheet of ice. Such a surface is the most slippery possible and the braking action is practically zero. The time necessary for completely dissolving the ice depends not only on the kind and concentration of the chemical but also on the meteorological conditions, thermal condition of the runway and the thickness of the ice layer.

Chlorides cannot be used for de-icing of airport movement areas because of corrosion, although they are cheap and effective. Sodium chloride may be used mixed with crushed gravel for access roads in some countries. If it is applied in its solid state and particularly if applied liberally, sodium chloride damages the pavement surface.

One of the most used chemicals is urea. The technical term urea is used for the amid of the carbonic acid – carbamid with chemical composition $CO(NH_2)_2$. Urea is a non-toxic substance, and its shelf life is unlimited.

The theoretical efficiency of urea is high up to -11.5 °C, the eutectic temperature, at 24.5 % concentration of the solution. Practically it is possible to use urea up to -5 °C. Its advantage is that it can be applied as a de-icing or as a preventive anti-icing chemical. Urea can be applied in the form of a solution, preferably warm, or by sprinkling granules. Sprinkling can be performed dry or the granules can be mixed with water and sprayed. Before application of urea as a de-icing fluid, it is necessary to clean the runway as much as possible by mechanical equipment. The spraying will take effect about 30 to 60 minutes after application. After the ice has softened it is necessary to complete the cleaning of the runway mechanically.

Table 12-2   Recommended concentrations of urea

Urea application	Urea solution [l m^{-2}]	Urea granules [g.m^{-2}]
Preventive	0.05 – 0.1	15 – 20
De-icing	0.15 – 0.35	30 - 70

It is better to use urea as a preventive chemical in a 35 % solution, which is safe against recrystallising up to – 8 °C. When used in this way for anti-icing, the urea concentrations are lower, as shown in Table 12-2. This results not only in lower costs but also in smaller impact on the environment, as discussed later

Higher doses of urea for de-icing pose a greater threat to the surface of the runway. This is particularly so for concrete surfaces. After the ice layer has been sprinkled with urea granules, a system with two components – **k** (CO(NH$_2$)$_2$ and H$_2$O) in three phases – **f** (urea, ice and water solution) will be formed, as shown in Figure12-6.

According to Gibbs phase rule, which determines the degree of freedom of system **v**, i.e. number of dimensions determining the condition of the system (temperature, pressure and composition), this is a monovariant system, i.e. the degree of freedom is only 1 according to the equation:

$$v = k + 2 - f$$

where:

v   number of degrees of freedom

k   number of components

f   number of phases.

The magnitude of the degree of freedom is the same as the degree of freedom of a eutectic mix. This means that under the given pressure and temperature this system is not in balance. Therefore the ice will start thawing and its temperature as well as the temperature of the runway surface will start decreasing as a result of the consumption of the latent heat of thawing. In the growing water volume, more urea will be thawed. This will continue to happen until the temperature of the mix decreases to the eutectic temperature. Of course, with a concentration of urea less than eutectic, the achieved temperature will be higher. This process leads to rapid cooling of the runway surface and to large differences in temperature inside its upper layer. This results in high internal tensions, which cause peeling of the top 3 to 5 mm of the concrete runway. If urea is applied in a warm solution the necessary thermal energy will partially be taken from the solution and the surface of the runway is not so stressed as in granular application.

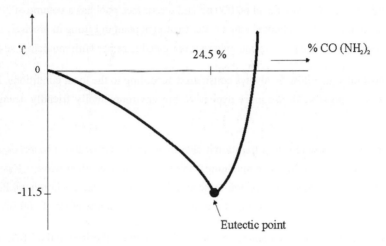

Figure 12-6   Ice - urea de-icing system

This peeling occurs particularly with new rigid runways, the surface of which is absorptive. At temperatures slightly below zero the water in capillary pores of the runway surface is not frozen. The rapid cool down as a result of application of the urea causes the water in the capillaries to freeze and this causes a further increase of stress in the surface layer of the runway. Therefore before winter, new runways have to be impregnated with a protective coating, which reduces the penetration of water into the surface.

Urea was originally used as a fertiliser. Nitrogen accounts for about 45 % of the total weight of urea. Application of urea increases the concentration of nitrates in ground and surface waters. There is then significant growth of algae in the watercourses and this disrupts the ecosystem, particularly where it is not possible to achieve sufficient dilution when the waste waters are discharged. Therefore the use of urea is forbidden or limited in some airports. The concentration of nitrates is often monitored in several parts of the drainage system of the airport and in outlets into watercourses. Limiting values of contamination of waste waters of NH_4 max. 3.5 mg.l^{-1}, NO_3 max. 15.0 mg.l^{-1} and dissolved materials 300 mg l^{-1} should be met when checked three to five hours after application of de-icing chemicals on the airport.

In order not to exceed the appropriate limits of nitrate concentration in surface waters, retention pools have been built on airports, in which contaminated waters are trapped and then gradually and slowly discharged. Alternative solutions are the construction of a water treatment plant or the combination of both solutions. The volume of waste waters from two runways at Munich airport is so large that the considerable capacity of a water treatment plant would not be utilised in the periods when the ice is not melted. Therefore two retention pools have been built on the airport. The underground tank has a volume of 60 000 m^3 and the surface pool has a volume of 20 000 m^3. Water from retention pools is admitted into a water treatment plant in Eitting in precisely specified volumes depending on the concentration, where it is treated together with municipal sewage.

The use of urea must therefore be always considered according to the local conditions. If its use is unsuitable, it is possible to use more expensive but environmentally friendly acetate-based chemicals.

Airport operators are now rejecting traditional chemicals in favour of those based on acetates. These are more effective at lower temperatures, they act for longer, they leave a less slippery surface, their storage is easier, they are non-toxic and do not damage the environment. Potassium acetate-based de-icing fluids are theoretically effective up to the temperature of –60 °C.

Acetate-based chemicals are recommended by the environmental authorities in the USA, in several countries of Western Europe and in Scandinavia. They are easily biodegradable with a small consumption of free oxygen in water. When they decompose, carbon dioxide and water are generated. They are, however, more expensive than the more traditional de-icers. It is probable that for those airports that have already built expensive facilities for the treatment of waste water from movement areas of the airport or retention pools as at Manchester, it will be more advantageous to carry on using urea or glycol, which are cheaper in comparison to the acetates. Because acetates have lower viscosity, they require different spraying devices than e.g. glycol, which means another additional investment for many airports. Therefore each airport must consider the advantages and disadvantages of individual chemicals in its specific situation.

Potassium acetate-based chemicals are used in solution for anti-icing. For runway de-icing it is better to use a combination of granules of sodium acetate and a solution of potassium acetate. The granules of sodium acetate are applied first. As the granules dissolve, heat is released and this creates holes in the ice. At the moment the solution of potassium acetate is applied, it will penetrate through the holes to the surface of the runway, where it acts from below and releases the ice form the surface of the runway.

Glycol-based de-icing chemicals have two disadvantages. Ethylene glycol and diethylene glycol are toxic. Moreover, ethylene glycol is considered to be carcinogenic. This can be overcome by using propylene glycol-based chemicals, which is considered non-toxic in many countries. Despite this, glycol is used on its own or in combination with urea because of its easy manipulation, storage and the price.

The disadvantage of all kinds of glycol is that upon decomposition in the water they consume a high volume of oxygen and this seriously endangers life in the watercourse. Therefore waste waters must be treated before being discharged into the watercourse. On Munich airport, glycol-based de-icing chemicals are used for de-icing taxiways. Treatment of waste waters and disposal of glycol is performed directly in special beds made of bentonite powder and sand built along the taxiways through which the waste waters seep. Bacteria are inoculated into the beds, which decompose glycol into water and carbon dioxide.

Ice can also be removed from a runway with high pressure water. The Küppel-Weisel company offers this technology. The water is sprayed under high pressure through nozzles against the runway. It penetrates through the layer of compacted snow or ice, disturbs it and separates it from the surface of the runway. Then the runway is mechanically cleaned. It is then treated by a small volume of glycol or by another de-icing chemical to prevent further icing.

12.5 THERMAL DE-ICING

Thermal procedures for removing of snow and ice are not so widely used as use of mechanical means and chemicals. The reasons are the growing price of energy and the problems with maintenance of some types of facilities. None-the-less, it is possible that in some airports the local conditions can be favourable, e.g. the utilisation of waste heat or geothermal energy for removing of snow and ice from runways.

Greater use of thermal methods, particularly heated runways, can be expected as the cost of human labour rises in the future. The advantage, particularly for large airports, is that the heated runway can be used constantly and does not have to be closed down to perform winter maintenance. Heating can be electric or by a heated pipe which is buried in the runway. The disadvantages of heated runways are high investment and operational costs and the need to maintain the system. The heating is switched on before it starts snowing or if there is a forecast of icing. The runway temperature is then maintained above the freezing point.

Melting of snow is hardly used any more because of high energy consumption and low efficiency.

Also at most airports the use of jet blowers, jet engines fixed on the undercarriage of the truck, is forbidden. They are expensive due to the price of fuel, noise and air pollution, but also because of the damage to the runway. Thermal de-icing can be effective only if not only ice will be melted but also the surface of the runway will be completely dried. The use of a jet blower causes an increase of local runway temperature, which will cause high internal stress in the concrete slab and the later appearance of cracks. The combination of high temperature and kinetic effects of jet exhaust gases can cause the destruction of the covering layer of asphalt runways.

12.6 RUNWAY SURFACE MONITORING

According to the requirements of Annex 14, the surface of the runway must be maintained in a clean state to ensure good braking action. However, from the point of view of protecting the environment and for economy, it is necessary to minimise the volume of chemicals used for removing ice. The application procedure, the application time and the necessary volume of the chemicals are affected by a number of factors, the effect of which can be difficult to assess only on the basis of operational experience. For example, a sufficient amount of heat can be accumulated on the runway pavement after a sunny day so that the pavement does not freeze even during a wet and cold night. An optimum decision can be made on the basis of data from the monitoring system, which monitors meteorological conditions and the condition of the runway and enables a forecast of the condition of the runway surface to be made. Because the monitoring system allows a substantial decrease in the requirements for winter maintenance and in the volume of chemicals used by up to 70 %, this investment usually has a high rate of return.

The monitoring system must be capable not only of indicating the appearance of ice on the runway surface but also to forecast conditions leading to icing. Modern systems such as that produced by the Vaisala company use as input data the output from the sensors of the Milos automatic station, which monitors the air temperature, dew point temperature, amount and ceiling of clouds, speed of wind, precipitation, temperature of the surface of the pavement, temperature under the surface of the runway and the residues of the de-icing fluid on the runway surface. The system is capable of forecasting the runway condition for 24 hours ahead. It is possible to enter the weather forecast into the system manually, and in this way to obtain various alternative forecasts.

On the basis of the forecast the following can be estimated:

➤ whether the temperature of the runway surface will fall below 0 °C

➤ when the temperature of the runway surface will fall below 0 °C

➤ whether the surface of the runway will be wet when the temperature is below 0 °C

✈ how long the temperature of the runway surface will remain below 0 °C

✈ whether the residues of de-icing material from the last application on the surface of the runway will prevent the formation of ice.

On the basis of experience from Frankfurt airport, a monitoring system that is too complicated is more of a burden than a useful aid. In the case of Frankfurt, where the sensors have been placed at 34 locations, the system indicates frequent failures. The Vaisala system decreases the necessary number of sensors and at the same time optimises their location by mapping the runway surface with a thermovisual camera. On each surface there are critical 'cold' places, which freeze before other surfaces on the runway. The runway surface is scanned by the thermovisual camera under three different conditions. During a cloudless and windless night it is possible to see the maximum contrast between the 'cold' and 'warm' places on the runway. Average conditions are 5/8 cloud cover and a 2.5 to 10 m.s^{-1} wind speed. A wet night with more than 6/8 cloud cover with the cloud base of less than 600 m gives the smallest temperature differences between individual places of the runway. After evaluation of the mapping, the sensors are placed into 'cold' places and for comparison into places which, from the point of view of temperature, are the most stable. In this way it is possible to simplify the monitoring system, to increase its reliability and to decrease the construction costs.

12.7 AIRCRAFT DE-ICING

It is dangerous for an aircraft to take off while it is contaminated by snow or ice. The contamination reduces the available lift due to the premature separation of the airflow from the wing surface and from the increase of aerodynamic resistance and aircraft weight. Research has shown that even a layer 0.5 mm thick covering the whole upper area of the wing can decrease the maximum lift coefficient by up to 33 % and the stalling angle of attack from 13 % to 7 %. This means that even a small layer of ice is not 'insignificant'.

De-icing fluids must meet a number of requirements. They must not be corrosive, they must not damage materials and aircraft coatings and must be non-flammable. Two basic types of fluids are used.

Type I – de-icing fluids are used for removing ice from the aircraft surface. However they do not provide long term protection against re-icing of the surface. They are most used in the USA. They contain more than 90 % glycol, the remainder consisting of water, corrosion inhibitors and wetting agents. Inhibitors ensure protection of the airframe, and wetting agents ensure the even coverage of the whole surface. Before use the de-icing fluid, it must be diluted with water in order to ensure its maximum efficiency corresponding to the concentration and temperature of the eutectic point (see Figure 12-7). Depending on the kind of glycol, this concentration corresponds approximately to 40 % of water and 60 % of glycol.

Type II – protective (anti-icing) **fluids** can be used for removing ice from the aircraft surface but mainly to protect the aircraft from re-freezing again during taxing and waiting before take-off. Protective fluids are normally applied to the cleaned surface of the aircraft.

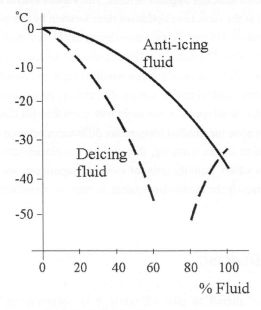

Figure 12-7   Comparison of the freezing point diagrams for aqueous solutions of Type I - de-icing fluids) against Type II - protective fluids

As with the Type I fluids, Type II fluids contain glycol, water, inhibitors and wetting agents. In addition to this they contain a polymer-based thickener. The thickener increases the viscosity of the protective fluid so that it adheres to the aircraft surface and does not flow away. The thickener is dissolved in water. The composition of protective fluids corresponds to minimum freezing point. The Type II fluids are not diluted any more.

Protective fluids are classed as Non-Newtonian, i.e. viscous fluids. The viscosity of a Non-Newtonian fluid changes depending on the shear on the fluid surface, as shown in Figure 12-8. The fluid has high viscosity at small values of shear and small viscosity if the shear on the fluid surface grows. This property is very important for protective fluids. When the aircraft is stationary or is taxiing, the fluid viscosity is high and the fluid adheres to the surface of the aircraft. During the take off run, the fluid experiences higher and higher shear, its viscosity is suddenly reduced and it flows from the surface of the aircraft. Ideally the fluid viscosity should be high up to a speed of about 110 km.h⁻¹ and then it should be reduced rapidly so that at the nose-wheel lifting speed, the aircraft surface is clean.

The two types of fluid are applied differently. First the aircraft is cleaned by the Type I. fluid then, within 3 minutes, the surface should be treated by the Type II fluid. To make the cleaning of the aircraft easier, the de-icing fluid is applied hot and under high pressure. At a temperature of up to −7°C some airports use hot pressurised water for cleaning of the aircraft. Protective fluids are less stable and must not be therefore exposed to higher temperatures. They are applied cold. In order to avoid disrupting the function of the polymerising fluid and reduction of the fluid viscosity, the protective fluid must not be pumped by common types of rotary or piston pumps. Instead, membrane or screw pumps must be used.

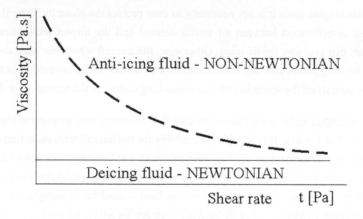

Figure 12-8   Viscosity against shear rate profiles for Newtonian and non-Newtonian fluids

De-icing of aircraft can be performed in either centralised or decentralised ways. The latter is by mobile vehicles on the apron and the former way is to use one or several specially constructed stands located as close as possible to the take-off runway threshold.

Type I and II fluids must be used in decentralised systems so that, during taxiing to take-off, the aircraft does not freeze again. The residues of de-icing fluids remain on the apron, and they penetrate into the substratum, where they contaminate soil and ground water. The residuals of the Type II. fluid together with the residuals of snow are very slippery, which endangers the safety on the apron. This cannot be avoided completely even on airports where de-icing is performed centrally, because propellers and the inlets of jet engines must be de-iced on the apron stands.

To ensure the protection of the environment, and to increase the safety of air operation by shortening the time between de-icing of the aircraft and its take-off, areas for central de-icing of aircraft are being built on the apron at larger airports. The design of the surfaces prevents the escape of the de-icing substance. Glycol is trapped and recycled. Recycling is in most cases performed in the production plant, because the recycled fluid must have the same properties as a new fluid, but e.g. in the case of the Munich airport recycling is performed directly on the airport. It is, in fact, difficult to trap and recycle much more than 50 % of the fluid. For central de-icing, it is possible to use mobile de-icing facilities or a special portal de-icing facility. Portal type de-icing facilities have been constructed on Luleå (Sweden) and Munich airports. Their advantage is substantially higher capacity, which is about 10 to 15 aircraft an hour in comparison with 4 to 12 per hour for de-icing by one mobile unit.

Another advantage of centralised de-icing is that de-icing can be performed immediately before take-off so that in most cases it is not necessary to then protect the plane by Type II. fluid. This requires close co-ordination between air traffic control and the airport administration or the handling agent that provides the de-icing. Otherwise, the aircraft which has been de-iced might have to wait too long before take-off. Co-ordination with the air-traffic control is just as important in de-icing of aircraft on the apron in order to avoid long delays to the aircraft after de-icing.

On Copenhagen airport and in some airports in the USA, the preventive spraying of planes by Type II. fluid is used after landing. It is suitable particularly for the aircraft with short turn-round time. Experience shows that as many as 95 % of aircraft which had been anti-iced after landing do not have to be de-iced before take-off. The advantage of anti-icing is substantially lower consumption of de-icing fluid. On average, 230 litres of de-icing fluid is used for de-icing an MD-80 aircraft. In contrast, less than 10 litres of Type II. fluid is necessary for anti-icing the same type of aircraft. There may also be an increase in airport capacity. Anti-icing takes less than a minute in comparison to between two and five minutes for de-icing. Even if some anti-iced aircraft must be de-iced before take-off, the consumption of de-icing fluid is still lower and the time is shorter than for aircraft that have not been anti-iced after landing.

De-icing of aircraft is a professional and specialised activity which has a direct effect on the safety of air transport. The workers who perform de-icing must be perfectly familiar with the de-icing equipment and with the properties of the fluids used. Increased attention has to be paid to the protection of these workers, both because of adverse weather conditions, during which the de-icing of aircraft is most frequent and necessary, and because glycol is harmful to the health.

13

AIRPORT EMERGENCY SERVICES

Tony Kazda and Bob Caves

Motto: Know safety – no pain; No safety - know pain.

13.1 ROLES OF THE RESCUE AND FIRE FIGHTING SERVICE

Airport administrations spend millions of dollars providing and operating fire fighting equipment, hoping that they will never have to use it.. A Rescue and Fire Fighting Service (RFFS) produces costs without explicit benefits for airport administrations. As with other departments that generate no revenue, when there is a squeeze on finances at an airport, management tends to minimise spending on the RFFS. From a long-term viewpoint, the airport image may suffer from such an attitude. The safety standard of an airport is high on a list of priorities of airlines. As a rule of thumb, in the event of a serious aircraft fire, each fireman might expect to save up to five passengers. If the RFFS is under-funded, the airlines may decide to fly elsewhere. From that viewpoint, the aerodrome RFFS is one of the inseparable parts that go to create the complete 'airport product'. If the services are of high quality, they may be an important factor in the marketing campaign of the airport.

Air transport is one of the safest means of transport. None-the-less, there is a finite probability that an accident will occur sooner or later. The accident statistics show that the majority of aircraft accidents happen during take-off and landing, that is to say, in the vicinity of airports. The airport operator must be prepared for such an eventuality. During take-off, there is a maximum quantity of highly inflammable fuel on board. A severe accident usually causes a break in the integrity of the aircraft's fuel tanks, leading to the escape and atomising of the fuel and a subsequent fire. An extensive fuel fire brings the greatest probability that the passengers will not survive the accident. The principal task of the RFFS is to save human lives. The necessary level of provision of RFFS does not depend only on the aerodrome size, for example with regards to the runway length to be covered in order to reach an accident site, but also on the size of the aircraft. The demands upon the RFFS at international airports have been gradually increasing because ever larger aeroplanes with greater fuel volumes and greater seat capacity have been put into operation.

Fortunately, serious accidents with fire, and hence a heavy death toll, occur only rarely (see Table 2-2). More often, there are smaller incidents like spilled fuel and incipient emergency situations like undercarriage failures that the RFFS must attend. Most events are not connected with aircraft at all, but concern fire alerts in the terminals, road accidents, and other emergencies that would be expected in any city.

13.2 LEVEL OF PROTECTION REQUIRED

13.2.1 Response Times

The required level of provision of the RFFS is established in Annex 14 and in Airport Services Manual: Part 7 Rescue and Fire Fighting.

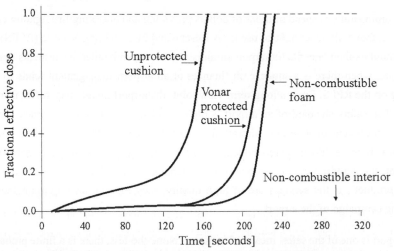

Figure 13-1   Effect of cushioning protection and materials on calculated survival time

The most important requirements are the speed and the effectiveness of the response to the emergency, which are defined by maximum times for the response to the emergency and to control the fire.

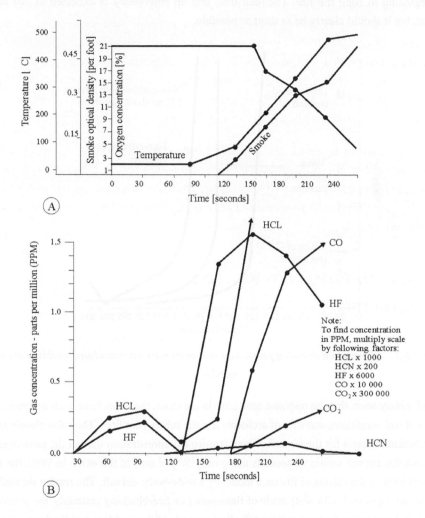

Figure 13-2   Aft cabin hazards resulting from burning interior materials

The maximum response time criterion is that the RFFS should reach any location in the runway system and any other part of the movement area where the accident may have occurred and begin the rescue operation preferably within two minutes and not exceeding three minutes in optimum visibility and surface conditions. The response time is the time between the initial call to the RFFS and the time when the first emergency vehicle is in position to discharge the extinguishing agent with at least 50 % of its maximum rate.

The control time criterion is that 90 % of the fire's extent should be extinguished within one minute from beginning to fight the fire. The total time that an emergency is expected to last is not specified, but it should clearly be as short as possible.

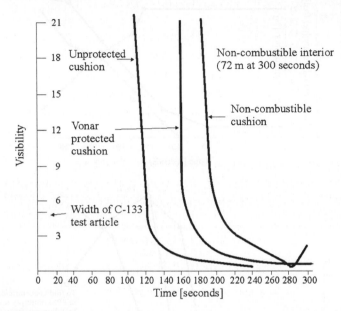

Figure 13-3   Effect of cushioning protection and materials on calculated visibility through smoke

Although it may seem that the response time criteria are strict, the times have been determined on the basis of real conditions, statistics of accidents and the subsequent fires. They also closely match the certification criteria for the aircraft, which require a demonstration that all the passengers be able to exit the aircraft within 90 seconds using only 50 per cent of the exits. In 1980, the FAA performed tests of the effects of fire on a model of a wide-body aircraft. The results showed that in aircraft not equipped with seats made of flameproof or fire-blocking materials, the possibility of rescuing the people who have survived the first seconds of the accident rapidly drops after 120 seconds from the beginning of the fire. If the seat upholstery is covered with a material called Vonar, or if the seats are entirely flameproof, the time of survival increases to approximately 180 seconds, as shown in Figure 13-1.

The flames themselves are not the only threat to life. The high temperature raises the concentration of asphyxiant and irritant gases, and lowers the level of oxygen in the cabin (see Figure 13-2). The evacuation of the passengers is aggravated by a rapidly declining visibility as a consequence of thick smoke. In some cases the visibility declines from the full cabin length to only the cabin width within 15 seconds (see Figure 13-3).

13.2.2 Aerodrome Category For Rescue And Fire Fighting

The required level of fire protection to be provided at an aerodrome should be determined by the dimensions and number of movements of the aeroplanes using the aerodrome. The aerodrome shall be ranked into one of ten categories based on the fuselage length and width of the longest (critical) aeroplane (see Table 13-1). If the width of the critical aeroplane fuselage is greater than the maximum allowed in Table 13–1, after selecting the category appropriate to the longest aeroplane, then the actual category shall be one number higher.

Table 13-1   Aerodrome category for rescue and fire fighting

Aerodrome category	Aeroplane overall length	Maximum fuselage width
1	0 m up to but not including 9 m	2 m
2	9 m up to but not including 12 m	2 m
3	12 m up to but not including18 m	3 m
4	18 m up to but not including 24 m	4 m
5	24 m up to but not including 28 m	4 m
6	28 m up to but not including 39 m	5 m
7	39 m up to but not including 49 m	5 m
8	49 m up to but not including 61 m	7 m
9	61 m up to but not including 76 m	7 m
10	76 m up to but not including 90 m	8 m

If the number of movements of aeroplanes included in the highest category for rescue and fire fighting during the three most busy consecutive months reaches or is greater then 700, the aerodrome shall be classified into the respective category.

If the number of movements of aeroplanes included in the highest category for rescue and fire fighting during the three busiest consecutive months is smaller then 700, the aerodrome category may be reduced by one grade. From 1 January 2005 it will not be possible to downgrade the aerodrome rescue and fire fighting category and the aerodrome classification should be equal to the respective category.

During low traffic periods, the level of rescue and fire fighting may be reduced so that it corresponds to those types of aeroplanes which are intended to use the aerodrome at that time.

The amounts of extinguishing agents or the amounts of water necessary for their preparation (see Table 13-3) and the numbers of fire fighting vehicles (Table 13-4) at the aerodrome shall be determined by the aerodrome's rescue and fire fighting category.

13.2.3 Principal Extinguishing Agents

Both principal and complementary extinguishing agents should be provided at an aerodrome. The only exception shall be aerodromes ranked in the 1st and 2nd category, where the entire amount of principal extinguishing agents may be replaced by complementary agents. The principal extinguishing agents facilitate the extinguishing and long-lasting control of fire.

The traditional principal extinguishing agent is a protein foam. It creates a resistant and stable foam cover, suitable for covering, extinguishing and cooling the hot aeroplane fuselage. The basic component of the foam concentrate is a hydrolysed protein complemented with stabilising additives suppressing decomposition of the foam at high temperatures, inhibitors of corrosion, antibacterial additives and antifreeze additives. It shall be mixed with water in a proportion 3; 5 % to 6 % according to the type of extinguishing device and the kind of extinguishing agent and its performance criteria. A problem may arise with incompatibility of the protein foam with some kinds of extinguishing powders which may decompose it. It usually complies with the criteria for the 'A' level principal extinguishing agents.

The Aqueous Forming Foam (AFFF) - is very effective for spilled burning fuel. It is a surface active agent with a foam stabiliser. The AFFF does not have the appearance of a protein or fluoroprotein foam. At high temperatures the layer of this agent decomposes. It is unstable. The AFFF is not suitable for fires containing large quantities of hot metal. The concentrate shall be mixed with water in concentration of 1 % to 6 %. The AFFF usually complies with the criteria for the 'B' level principal extinguishing agents.

The fluoroprotein foam is a combination of the protein foam with fluorocarbonate. It is very effective at extinguishing deeper layers of fuel, for example for subsurface application in fuel tanks, because it inhibits penetration of the burning fuel to the surface through individual foam bubbles. In comparison with the protein foams, it demonstrates better compatibility with extinguishing powders. Although the fluoroprotein foams are generally more expensive than the protein ones, their advantage is in higher efficiency and lower consumption of water. Depending on the type of extinguishing device, it shall be mixed with water in a proportion 3 % to 6 %. It is an agent complying with the criteria for the 'B' level principal extinguishing agents.

The Film Forming Fluoroprotein Foam (FFFP) is a modern extinguishing agent, highly effective for spilled fuel. It is a combination of protein foam with a surface active agent on a base of fluorocarbonate. The extinguishing agent forms a water solution with properties of an oily material on the surface of spilled fuel. Therefore the extinguishing agent rapidly spreads on the burning fuel surface, inhibiting the access of air and suppressing evaporation of the fuel. As with other agents, it is used in a concentration of 3 % to 6 %. It is compatible with the majority of extinguishing agents. It complies with the criteria for the 'B' level principal extinguishing agents.

Table 13-2   Foam specifications

Fire tests	Performance level A	Performance level B
1. Nozzle (air aspirated)		
a) Branch pipe	UNI 86 – foam nozzle	UNI 86 – foam nozzle
b) Nozzle pressure	700 kPa	700 kPa
c) Application rate	4.1 l/min/m^2	2.5 l/min/m^2
d) Discharge rate	11.4 l/min	11.4 l/min
2. Fire size	2.8 m^2 (circular)	4.5 m^2 (circular)
3. Fuel (on water substrate)	Kerosene	Kerosene
4. Preburn time	60 s	60 s
5. Fire performance		
a) Extinguishing time	≤ 60 s	≤ 60 s
b) Total application time	120 s	120 s
c) 25 % re-ignition time	≥ 5 min	≥ 5 min

The synthetic foams are produced from petroleum products. They are convenient for extinguishing fires in closed rooms and to some extent for spilled fuel.

The specifications on 'A' and 'B' principal extinguishing agents include physical properties and the performance of the foams under fire test conditions. These include the data on the discharge monitors, quantity and velocity of delivery of the extinguishing agent, the extent of the fire, the speed of extinguishing the fire, and on the resistance of the extinguishing agent against a new flare-up of the fire (see Table 13-2).

If it is necessary to raise the aerodrome category for rescue and fire fighting, and an extinguishing agent complying with the 'A' level criteria is used at the aerodrome, the agent may be replaced by an extinguishing agent complying with the 'B'level criteria which requires smaller quantities of water.

13.2.4 Complementary Extinguishing Agents

The complementary extinguishing agents should be dry chemical powders, halons, and carbon dioxide. The complementary extinguishing agents are very effective and facilitate quick suppressing of the fire. However, they are effective only while being applied, and do not protect against a new flare-up of the fire. Therefore they are used for the first strike, before the arrival and intervention of the major fire engines. They are also effective within enclosed rooms like baggage lockers, spaces under the aeroplane wings and undercarriage bays, where the foams are not effective. Their further use is during special small fires, such as to tires, wheel discs, engine, etc.

Dry chemical powders are among the most effective complementary extinguishing agents. They are the most convenient agents for extinguishing fuel spills and three dimensional fires. A disadvantage of the powders is that they do not cool the extinguished object when they are applied. If a greater amount of powder is used, it may form a cloud which reduces visibility and causes breathing difficulties. Powders are based on potassium hydrocarbonate, potassium chloride and ammonium phosphate. Carbamide powder, a product of the reaction of potassium hydrocarbonate and urea and offered in the market under the name Monnex, is several times more effective. Dry powders should comply with the ISO 7201 standard.

An advantage of carbon dioxide is that during the intervention it cools the material at the same time. After its application, it is not necessary to clean the equipment. It is an effective agent for small fires in closed rooms. Carbon dioxide should comply with the ISO 5923 standard.

An Amendment of Annex 14, Volume I was proposed in March 1999 that only dry powders be used as the complementary agents. Halon production has been banned since 1 January 1994 and there is an increasing concern about the global warming potential of carbon dioxide and the effect of halogenated carbons on the depletion of the ozone layer in the atmosphere. Carbon dioxide is generally not used due to the bulkiness of the containers.

13.2.5 The Amounts of Extinguishing Agents

The amount of water for particular aerodrome categories shall be derived from a theoretical critical area which should be extinguished. The theoretical critical area is an area around the aeroplane in which the fire should be extinguished, and which allows survival and escape of the passengers from the aeroplane. The dimensions of the critical area have been determined experimentally. The theoretical critical area has a rectangular shape. One of its sides is equal to the length of the aeroplane fuselage. The other side of the critical area is equal to the greatest width of the aeroplane augmented by 24 m against the wind and 6 metres downwind for aeroplanes whose fuselage length is equal to or greater than 20 m. For aeroplanes with a fuselage length shorter than 20 m, the critical area shall be widened by 6 m on both sides from the greatest width of the aeroplane fuselage.

Table 13-3   Minimum usable amounts of extinguishing agents

Aerodrome category	Foam meeting performance level A		Foam meeting performance level B		Complementary agents or or		
	Water [1]	Discharge rate foam solution{ minute [1]	Water [1]	Discharge rate foam solution{ minute [1]	Dry chemical powders [kg]	Halons [kg]	CO_2 [kg]
1	350	350	230	230	45	45	90
2	1 000	800	670	550	90	90	180
3	1 800	1 300	1 200	900	135	135	270
4	3 600	2 600	2 400	1 800	135	135	270
5	8 100	4 500	5 400	3 000	180	180	360
6	11 800	6 00	7 900	4 000	225	225	450
7	18 200	7 900	12 100	5 300	225	225	450
8	27 300	10 800	18 200	7 200	450	450	900
9	36 400	13 500	24 300	9 000	450	450	900
10	48 200	16 600	32 300	11 200	450	450	900

In the majority of incidents, only two thirds of the theoretical area - the real critical area - are caught by the fire. The total amount of water shall be determined from the real critical area, the amount of water required per m^2 per minute and the extinguishing time, with 60 s as the control time. Then the amount is augmented by the water required for extinguishing the remains of the fire. The minimum amounts of water for extinguishing agent production and particular aerodrome categories are given in the Table 13-3.

In many cases, considerably larger amounts of water are required for extinguishing an aeroplane fire than those recommended in Table 13-3. In the majority of events the amount may even double. Therefore it may be expected that in the future the requirements for the amount of water will be raised.

It must be emphasised that the determined minimum amounts of extinguishing agents should be constantly available on rescue and fire fighting vehicles. In the event that all the rescue and fire fighting vehicles participate in the intervention, the aerodrome should be closed. If only a certain number of the rescue and fire fighting vehicles participate in the intervention, the aerodrome category for rescue and fire fighting should be temporarily downgraded. The administrations of large airports make efforts to provide a continuous operation, and if possible, without any limitations. In that case they must provide a greater quantity of extinguishing agents and numbers of rescue and fire fighting vehicles than required for a single emergency.

13.3 RESCUE AND FIRE FIGHTING VEHICLES

The task of the rescue and fire fighting vehicles is to reach the accident location as soon as possible, to provide escape routes, to liquidate the fire, and to begin rescue actions.

Table 13-4   Number of rescue and fire fighting vehicles

Aerodrome category	Rescue and fire fighting vehicles
1 – 5	1
6 – 7	2
8 – 10	3

The minimum determined numbers of rescue and fire fighting vehicles must facilitate the transport of the required amounts of extinguishing agents to the fire's location and their effective application. The number of rescue and fire fighting vehicles depend on the aerodrome category, as shown in

Table 13-4. In determining the number of rescue and fire fighting vehicles, not only the amount of extinguishing agents carried by them is important, but also the possibility of applying them so that the entire aeroplane fuselage can be covered. If, for example, 3 600 litres of extinguishing agent should be provided at the aerodrome, is it more appropriate to have available two vehicles, each of them carrying 1 800 litres, with which the attack may be conducted from two points, than only one with the entire amount. At the same time, it is necessary to assess the capital cost of the rescue and fire fighting vehicle, and their replacement approximately every 10 to 15 years, and the operating costs, which will be higher with two vehicles. The capacity of the vehicle tank for foam concentrate should correspond to at least two full loads of such quantity of water that is necessary for foam production at the determined concentration.

The original ICAO concept determined separately the numbers of the so called rapid intervention vehicles equipped in the majority of cases with complementary agents, powders in particular, and major vehicles which carried the load of water for the preparation of principal extinguishing agents. The rapid intervention vehicles reached the accident location first and began the fire fighting intervention.

The major vehicles had to reach the incident location before the rapid intervention vehicle consumed all its extinguishing agents. That concept was invalidated by the technical limitations of the older types of major vehicles with which it was not possible to fulfil the required response time. Now, parameters are determined separately for vehicles with a volume of water up to 4 500 l and above 4 500 l (see Table 13-5). The fire fighting intervention should be started within the limit of the response time with at least 50 % of the performance given in the Table 13-3. All the remaining vehicles must reach the accident location within 1 minute after the first vehicle.

The majority of vehicles are custom-made in close co-operation with the customer. Besides the basic equipment, the customer may order other special equipment which exceeds the current standard. In an attempt to reduce the costs of RFFS, personnel costs in particular, efforts have been made to use more and more sophisticated rescue and fire fighting vehicles, so reducing the number of employees. Such equipment may substantially ease the intervention and make it more effective, or minimise human factor errors. On the other hand, it increases costs of acquisition and maintenance. The failure of any of the sophisticated components might cause the vehicle to be unavailable. If a failure occurs of the main systems of the rescue and fire fighting vehicle during a rescue and fire fighting intervention, for example the monitor or valves which are equipped with a remote control with boosters, it should be possible to disable the remote control and to control the system manually. Therefore, when specifying the requirements for a new vehicle, a compromise must be made between the vehicle's simplicity, the extent of the additional equipment, price, operating costs, and its availability.

Table 13-5 Minimum characteristics recommended for rescue and fire fighting (RFF) vehicles

Parameter/equipment	RFF vehicles up to 4 500 l	RFF vehicles over 4 500 l
Monitor	Optional for categories 1 and 2 Required for categories 3 to 10	Required
Design feature	High discharge capacity	High and low discharge capacity
Range of monitor	Appropriate to longest aeroplane	Appropriate to longest aeroplane
Handiness	Required	Required
Under truck nozzles	Optional	Required
Bumper turret	Optional	Optional
Acceleration	80 km.h^{-1} within 25 s	80 km.h^{-1} within 40 s
Top speed	At least 105 km.h^{-1}	At least 100 km.h^{-1}
All-wheel drive	Yes	Required
Automatic or semi-automatic transmission	Yes	Required
Single rear wheel configuration	Preferable for categories 1 + 2 Required for categories 3 to 10	Required
Minimum angle of approach and departure	30°	30°
Minimum angle of tilt (static)	30°	28°

The assessed optional characteristics and equipment of the vehicle may be classified into two basic groups:

✈ fire fighting and rescue equipment

✈ performance and driveability of the vehicle.

One of the basic requirements is a perfect view from the vehicle facilitating an effective performance of the rescue and fire fighting intervention even in bad visibility. Nozzles that spray the extinguishing agent under the vehicle may permit the vehicle to pass through the fire, though some would argue that the fire should have been knocked down before the vehicle penetrates the area. The monitor on the front bumper facilitates fighting a fire in such spots that are beyond the reach of the main monitor installed on the roof of the vehicle, for example under the aeroplane wing. If the aerodrome is used by aeroplanes with rear mounted engines, or with the third engine installed in the fin of the aeroplane, it is necessary to be able to apply extinguishing agents up to a height of 10.5 m.

It should be emphasised that during each rescue and fire fighting intervention the firemen act under a great psychological pressure, sometimes overcome with stress to the extent that they need professional support after the emergency. Although the modern vehicles permit a single fireman to perform the rescue and fire fighting intervention, it is more convenient if one fireman is dedicated to driving and the other to the control the monitor. The cabin should protect the crew completely. A low level of noise allows communication between the crew members and by the means of a radio with the air traffic, or other staff participating in the rescue and fire fighting intervention. The rugged front of the vehicle should facilitate its passage through obstacles, such as gates, fences and smaller trees. Ventilation or air-conditioning in the cabin is important as protection of the crew against heat. It should be possible to hold the control handles and levers with hands in gloves, and the symbols must be readable even in bad light conditions and under vibrations.

The majority of manufacturers of the rescue and fire fighting vehicles mount the special fire fighting equipment on different types of undercarriage. Thus the buyer has the possibility of unifying the vehicle stock and of reducing the costs of spare parts.

In addition, when contemplating the purchase of a new rescue and fire fighting vehicle, it is necessary to consider its size and mass relative to the parameters of the communications and equipment in the area of the aerodrome and in its close vicinity. In particular, the dimensions of underpasses, tunnels, the load bearing capacities of bridges and pavements which may be used by the vehicle during a rescue and fire fighting intervention should be noted. In the majority of events, the vehicle needs to be capable of coping with poor terrain and thus have an all-wheel drive.

If in the aerodrome vicinity there is difficult terrain, special vehicles and equipment should be available. The terrain may include:

→ large water surfaces, in the area of approach and take-off in particular

→ swamps and similar areas, including river estuaries

→ mountain areas

→ deserts

✈   areas with large quantities of snow in winter.

Helicopters, hovercraft, boats, amphibians and land rovers may be utilised as special means and vehicles.

The RFFS requires a whole range of ancillary equipment if it is to function successfully in all circumstances. These include breathing apparatus and qualified staff to use it, extending ladders and floodlighting masts.

13.4 AIRPORT FIRE STATIONS

The location of a fire station at an airport is crucial for fulfilling the response time requirements. All the remaining criteria which may influence the location of the fire station among the other facilities in the airport are subordinate. In order that the response time limit can be fulfilled, two or more fire stations should be provided in some airports with an extensive runway system. Each of them should be located near the ends of the instrument runways, where accidents occur most. If more airport fire stations have been provided, one of them is usually established as the main fire station, and the remaining are designated as satellite sub-fire stations. One or more rescue and fire fighting vehicles are located in each airport fire station. The total number of extinguishing agents is divided so as to be sufficient for the first intervention. Vehicles from the second airport fire station should reach the place of accident within one minute after the first rescue and fire fighting vehicle.

When the location of a new fire station is considered, an assessment should also be made of the further development of the airport, and of the runway system in particular. If possible, the access from the fire station into the runway should be direct, without unnecessary turns.

In the past, airport fire stations were designed very functionally with minimum facilities for the crews. On the basis of investigations and operational experience, it has been found that the precondition for a successful rescue and fire fighting intervention is not only perfect technical equipment but also the psychological and physical well-being of the rescue crew. The extent and size of particular rooms should differ according to whether it is the main airport fire station or a satellite one, but in the majority of cases there will be the following facilities:

✈   garages for the rescue and fire fighting vehicles allowing their regular daily maintenance

✈   work and rest facilities for the staff

✈   communication and alarm systems

✈ storage rooms for technical support of the rescue and fire fighting vehicles, provision of extinguishing agents and fire fighting outfit and equipment.

The garages should be designed in such a way that there is enough space around the vehicles, at least 1.2 m on each side, even if bigger vehicles are acquired in the future. The garages should be heated to a temperature of 13 °C as a minimum. The rescue and fire fighting vehicles require connection with a supply of direct current for accumulator charging and permanent heating of oil, and compressed air. When the vehicle starts, the connectors must be automatically disconnected. In order that it should not be necessary to leave the garage with the vehicles to make the regular engine checks, ventilation should be provided for the exhaust gases. It must be possible to open the door quickly, preferably by a remote control from the watchtower. In the event of a failure of its automatic system, it should be possible to open it manually. A broken-down vehicle must not block the exit for the others.

The work and rest facilities for the staff should include locker rooms and sanitary installations, a multifunctional room for training and resting, with a small kitchen for the preparation of simple meals. The size of the offices will depend on the extent of tasks to be provided by the respective airport fire station.

A view should be provided over the largest portion possible of the movement areas from the watchtower. The alarm is usually given from the watchtower, and activity with other units co-ordinated from there during a rescue and fire fighting intervention. The communication systems of the watch tower should provide:

✈ direct telephone connection with air traffic control

✈ radio connection at the frequencies of the air traffic control for monitoring the radio correspondence, or for a direct communication with the crew of the aeroplane

✈ radio connection with the rescue and fire fighting vehicles, or a connection between the main and satellite airport fire stations (if constructed)

✈ telephone connection with fire fighting units in the airport vicinity and other services participating in a rescue and fire fighting intervention (police, hospital)

✈ a siren or other alarm system and a public address radio system.

The majority of airports are not prepared to receive the great number of telephone calls which follow a flight accident. For example, after the destruction of the B 747 Pan American air liner, Flight 103, over Lockerbie in December 1989, approximately 15 000 calls were registered every 15 minutes at Heathrow Airport. In the event of an accident to an aeroplane with 140-seats, on the first day, approximately 10 000 to 12 000 telephone calls may be expected at the airport. On the second day, that number drops by one half and on the third day by one half again. Thus the callers block the telephone exchange which is needed for the connection with other units according to the alarm schedule. There are computer systems available which allow the telephone lines to be used only for outward calls, and according to a programme, they dial sequentially the particular units that are supposed to participate in the rescue and fire fighting intervention. Provision should be made for the reception and briefing of the meeters and greeters who will need information on their relatives and friends.

Storage areas and other areas of technical support should allow the storing of extinguishing agents, their loading into vehicles and the storing and maintenance of rescue and fire fighting suits and equipment. A tower should be constructed for the drying of hoses and other equipment. The airport fire station should have its own stand-by source of electrical power or a connection to another stand by source. Emergency water tanks should be located near the runway thresholds to replenish the tenders.

13.5 EMERGENCY TRAINING AND ACTIVITY OF RESCUE AND FIRE FIGHTING UNIT

13.5.1 Training

The number and the level of training of firemen at the airport should correspond to the types of rescue and fire fighting vehicles used at the airport and to the maximum supply of the extinguishing agents for which the vehicles are designed. In determining the number of professional firemen, consideration should be given to the types and number of movements of the aeroplanes using the aerodrome. If there are only a few movements of aeroplanes during the day, the jobs may be done one after another by a single workman. For example, fireman and loading and unloading worker. All the personnel intended to participate in rescue and fire fighting actions should have specialised training and periodical tests in order to be able to perform their duties. In addition, the personnel who are intended to operate the rescue and fire fighting vehicles should be familiar with the driveability of the vehicle on different types of surfaces, including soft terrain.

Personnel should be trained so that, in the event of a rescue and fire fighting intervention, they are able to perform the activity entirely independently, without orders being given by the rescue commander. In critical situations, they should be capable of making decisions and to assume responsibility for their decisions.

The training of the rescue and fire fighting unit personnel may be divided into two categories:

+ basic training in skills and use of the rescue and fire fighting equipment, familiarisation with the airport vicinity and aircraft equipment

+ tactical training in extinguishing different types of fires and equipment, fire fighting and rescue actions.

The entire training programme should ensure that all members of the rescue and fire fighting unit are convinced that they are able to control fully the respective equipment, that the equipment functions without failures and that they are able to manage the assigned task. The acquired skills and abilities quickly degrade if not continuously practised.

The basic skill training comprises knowledge of different kinds of fire, types of extinguishing agents and their utilisation, technical parameters and control of all equipment to such a level that the usage and control of the equipment becomes fully automatic. An aspect of becoming familiar with the airport vicinity is the task of recognising both the most rapid and alternative routes to different areas under any type of weather conditions. All members of the rescue and fire fighting unit should, most importantly, know all the types of aircraft using the airport, in particular:

+ location of the exits and emergency exits of the aeroplanes, and how to open them

+ configuration of seats

+ kind and amount of fuel and location of tanks

+ location of accumulators

+ position of the points designed for a forcible entry into the aeroplane fuselage.

The purpose of the operational training is to perfect rescue and fire fighting behaviour using models simulating real conditions of an aeroplane and its parts, such as engines, undercarriages, aeroplane fuselage, cabin, and the like, on fire under different weather and visibility conditions. The training should be repeated in intervals shorter than one month.

Depending on the airport size and the national regulations, the airport operator is obliged to perform a full fire emergency exercise once a year, simulating the response to a high extensity aircraft accident. The purpose of such training is to verify the competency of the rescue and fire fighting unit and the co-ordination of activity with the units of fire services in the airport vicinity, police and health and emergency service. These exercises should be complemented by much more frequent 'table-top' exercises.

13.5.2 Preparation for an Emergency Situation and Rescue and Fire Fighting Intervention Control

At every airport, an emergency plan must be prepared for different types of emergency situations. The main purpose of the emergency plan is to minimise the effects of an unexpected event while also minimising the effect on flight operations. The airport emergency plan comprises the procedures which should be followed during the given type of emergency situation. The procedures include the sequence and range of responsibilities of each of the units participating in the rescue and fire fighting intervention. A serious accident always requires co-ordination and close co-operation with the fire units and hospitals in the airport vicinity. The units participating in the rescue and fire fighting intervention may include air traffic control, fire units of the airport, local fire brigades, police, emergency service, hospitals and health centres, telecommunications, military, civil aviation authority and representatives of the airline company. The main communications shall be between the airport fire unit and the air traffic control unit.

The contents of the airport emergency plan may be divided into:

1. Preparation of the airport for emergency situations, including the issues concerning organisational structure, responsibilities, testing the functional status of the equipment and clothing, and training of the personnel.

2. The activity during an emergency situation. It section describes procedures and tasks of the particular units participating in the rescue and fire fighting intervention.

3. The activity after termination of the emergency situation. These are activities which must be completed, but do not require an immediate solution, and the return of the airport to its regular operation.

The airport emergency plan should cover the issues concerning accidents both on the aerodrome and also outside the of the aerodrome boundary. The control of the rescue and fire fighting intervention at an accident on the aerodrome is usually within the competencies of the airport administration. The control of the rescue and fire fighting intervention at an accident out of the aerodrome depends on the laws and regulations in the given country and on the agreements between the airport administration and rescue and fire fighting units in the airport vicinity.

In some states the activity of the rescue and fire fighting service is limited to the area demarcated by the airport perimeter. In other countries, the area of responsibility is greater. When there is an accident of an aeroplane off the aerodrome, the necessary extent and the rapidity of the rescue and fire fighting intervention is, in the majority of cases, beyond the possibilities of local fire units. In addition, it is a specialised activity. Therefore the airport fire unit often responds to an accident a considerable distance from the airport. For example, at an emergency landing of SAS MD-81 airliner on 27th December 1991, 20 km north-west from Arlanda, three fire fighting vehicles were sent there despite the aerodrome category for rescue and fire fighting of the Stockholm - Arlanda airport having to be temporarily lowered. In the event of an accident more than 4 kilometres beyond the perimeter of London Heathrow Airport, the airport usually sends 50 % of its rescue and fire fighting units to the place of the accident. Further than that, a special vehicle with cutting and extrication devices will be probably sent at the request of the intervention commander.

Figure 13-4   The airplane recovery after runway end overrun, photo: J. Mach

The emergency plan should comprise the maps of the airport and its surroundings with grids in two different scales. The more detailed grid maps shall include access roads to the airport, location of water sources, emergency assembly points and parking bays. In addition to those data, the other map shall include information about hospital facilities to a distance of approximately 8 km from the airport reference point. By the means of a grid map, it is possible to identify with sufficient exactness the place of accident or of any other point in the area in question. In conditions of low visibility, the rescue and fire fighting unit may be guided to the area of the accident by means of radar in co-operation with the air traffic control. The officer in charge shall establish three zones of activity. Access to the crash zone should be confined to the RFFS. The rescue zone should be close by, and provide the immediate reception zone for survivors. The command post will normally be there, which will also be the rendezvous point for vehicles so that they can receive direct instructions if there is any sort of communications breakdown. The medical zone should be near hard standing, so that ambulances can drive up easily.

The activity of the rescue and fire fighting unit shall differ according to the category of emergency. The particular types are indicated as follows:

3rd degree of alert – **'aircraft accident'** - is declared if an aeroplane accident happens in the airport or in its vicinity. It shall be declared by the air traffic control or by the watchtower of the fire fighting unit.

2nd degree of alert – **'full emergency'** - is declared if it is known, or if there is a suspicion that an aeroplane on approach has problems which may lead to an accident. It is declared by the air traffic control.

1st degree of alert – **'local standby'** - is declared if it is known, or if there is a suspicion that an aeroplane in approach has a failure, however not of such a nature that it could seriously affect the safety during landing. The local standby is declared by the air traffic control.

13.6 RUNWAY FOAMING

Passenger aeroplanes with the mass over 5 700 kg with damaged or retracted undercarriages cannot land on a grass strip for emergency landing. On a strip with a low load bearing strength, the aeroplane would sink in and decelerate abruptly, with extensive damage to the structure and possible casualties. Therefore heavy passenger aeroplanes must use a paved runway. In these cases, the runway or a part thereof is sometimes covered with a protein foam blanket. Other kinds of extinguishing agents are not suitable for covering of the runway because they quickly decay.

In spite of the fact that the advantages of a foam blanket have not been entirely vindicated, it is recommended to lay one down if the aircraft captain is convinced that the foam will increase the level of safety during an emergency landing.

Theoretical advantages of a foam blanket are as follows:

✈ reduction of deceleration forces

✈ reduction of damage to the aeroplane

✈ reduction of the risk of creating sparks when the aircraft slides across the runway surface

✈ reduction of the risk of spilled fuel catching fire

Analysis of emergency landings of aeroplanes with retracted undercarriages has not fully confirmed the theoretical advantages of the foam blanket. Neither is there any evidence to prove that laying the blanket has any psychological advantages for the pilot.

When an assessment is made of the possibility of laying a foam blanket, besides the theoretical advantages, consideration should be given to a whole range of operational issues, either affecting the aircraft engineering, or the airport and its operational problems. As far as the aeroplane is concerned, the extent of damage to, or failure of, its undercarriage or other damage is important. Consideration should also be given to the pilot's experience, weather conditions, particularly the visibility, and the runway equipment. It is not always possible for the pilot to use a preferred runway, as shown by the accident at Amsterdam with the El Al Boeing 747 Freighter.

When an evaluation is made of the possibilities of the airport using the foam blanket, the factors include the number of runways, runway length on which the blanket is intended to be laid, traffic density, and consequences of closing of the runway for several hours or days, or cleaning and repair of the runway after the disabled aeroplane removal. Special equipment should be available for runway foaming. Taking into account its price and the rarity of such an event, not every airport carries a supply of foam. It is usually convenient to equip one airport in a given region. When a selection is made of which airport, besides the operational factors, consideration should be given also to the possibilities of the airport surroundings as far as the hospital capacities, approach communications, etc. are concerned. Extrication devices must be available, together with cranes and facilities to repair aeroplanes. Airports that do not have special equipment available should not try to provide runway foaming. The rescue and fire fighting vehicles should be constantly prepared for an intervention and have available the specified quantities of extinguishing agents. The captain of the aeroplane with the emergency may be forced by circumstances to execute the landing before runway foaming is finished, and the rescue and fire fighting vehicles should be prepared for this.

It is also necessary to assess the time which is available for foaming. Even with good organisation, the preparation and laying of the foam blanket will take one hour or more. If the aeroplane has enough fuel, it may fly off, reduce the mass and thus also the risk of a subsequent fire.

Weather conditions also have an effect upon the quality of the foam blanket. At high temperatures and overheated runways, the foam may decay more quickly. Conversely, at temperatures below zero, the water from the decaying foam freezes and considerably reduces the braking effect. It is not possible to lay the foam blanket in heavy rain or snow.

Foaming may be requested by the aircraft captain or by a representative of the air carrier. However, they should have sufficient flight and operating experience to consider in advance all these points, and also other factors, which have an effect on that decision.

Table 13-6   Water and protein foam liquid requirements for runway foaming

Parameter	Malfunction nose wheel	Wheels up landing			
		2-engine propeller	2-3 engine jet	4-engine propeller	4 engine jet
Width of pattern	8 m	12 m	12 m	23 m	23 m
Length of pattern	450 m	600 m	750 m	750 m	900 m
Runway area covered	3 600 m²	7 200 m²	9 000 m²	17 250 m²	20 700 m²
Water required	14 400 l	28 800 l	36 000 l	69 000 l	82 800 l
Protein foam required					
3 % type	432 l	864 l	1 080 l	2 070 l	2 484 l
6 % type	864 l	1 728 l	2 160 l	4 140 l	4 968 l

The touch-down point of an aeroplane with retracted undercarriage on the runway is, in the majority of cases, 'longer' by 150 m to 600 m than normally, as a consequence of a more significant 'ground effect'. The width, length and manner of runway foaming should correspond to the type of aircraft, to the kind of undercarriage failure and to the requirements of the pilot in command. In bad visibility, when it is difficult to distinguish where the blanket begins, it should be marked in an agreed way.

After foaming, it is necessary to leave the foam 'resting' for 10 to 15 minutes, so that the runway surface may dampen. However, the foam must not lie on the runway for a very long time, for example, more than 2 1/2 hours, particularly in summer, since water from the blanket runs out and the foam dries up. The foam blanket thickness should be approximately 5 cm. Table 13-6 gives water quantities required for the runway foaming.

13.7 POST EMERGENCY OPERATIONS

After the airport emergency is finished, the operations of the rescue and fire fighting units should be organised in such a way that evidence is not destroyed which may be pertinent to the subsequent investigation. Ambulances and rescue and fire fighting vehicles should not cross the track that the aeroplane left during the emergency landing, but they should, if possible, move along and to the side of it.

The state aviation authority should be immediately advised about the aeroplane accident, and a committee should be appointed to investigate it. Investigation of the accident usually requires several hours or even days during which a plan for the removal of the aeroplane may be established and the required equipment may be provided. Until the committee terminates the investigation and gives approval, the wreck of the aeroplane must not be moved. The only exceptions are when the wreck creates an obstacle and is a serious danger for the air traffic. If the wreck is blocking the runway end, consideration should be given to the possibility of operation on a shortened runway.

The disabled aeroplane usually disrupts the regular operation of the aerodrome, and therefore the airport administration requires its removal as soon as possible. The damaged aeroplane is the property of the flight operator and the insurance company. The airport administration may in no way touch the wreck, otherwise it would expose itself to the possibility that it will be accused of further damaging the aeroplane as a consequence of unauthorised and unprofessional manipulation of the aeroplane. Therefore the management of the disabled aircraft removal shall be in an exclusive competence of the flight operator and the insurance company, with whom the airport administration closely co-operates. The removal of small aeroplanes is a relatively simple matter. Upon agreement with the proprietor, the airport administration may undertake to do it.

The recovery of large passenger aeroplanes is technically and financially demanding. It is often dangerous, requires special equipment and knowledge in order that it should not be damaged. The airport administration usually has not the necessary equipment and mechanisms for such an activity. A specialist team shall be established for the removal of the disabled aeroplane, to which representatives of the operator, insurance company and manufacturer shall be appointed. The specialist group shall elaborate a plan for the removal of the aeroplane, which comprises in particular the following:

✈ a list of the required equipment, personnel and schedule of the activities

✈ preparation of the access roads for heavy-duty equipment

✈ surveying the place of accident

✈ safeguarding the place of accident

✈ manner of leasing the special equipment and mechanisms

✈ data and recommendations of the manufacturer

✈ defuelling, and draining of the oils and other liquids from the aeroplane.

Figure 13-5   Post emergency requires close co-operation with local companies who own heavy-duty mechanisms, photo J. Mach

In order that the disabled aeroplane can be removed, it is usually necessary to lift it, insert an emergency undercarriage under it, and then remove it on a temporary strengthened surface. Special air bags are most often used for lifting aeroplanes. The air bags are inserted under the wings, and the front and rear parts of the fuselage so that the aeroplane stability can constantly be ensured. Lifting is controlled by gradual inflation of the individual bags from a compressed air generator.

It is usually possible to lease the heavy-duty mechanisms required for removal of the aeroplane, such as cranes, lorries and machines for treatment of the access roads from local companies. The airport administration should have available the information about readily obtainable particular kinds of heavy-duty plant.

The other group shall be formed of special equipment usable for all types of aircraft. These are hydraulic lifts, air bags, compressors, lighting equipment, etc. Since it would be uneconomic to have such equipment at every airport, the IATA International Airlines Technical Pool programme was established, by means of which the special equipment is made available. The equipment may be used by the member airlines, and by the non member airlines for a charge. The equipment is kept at 11 airports (Bombay, Chicago, Honolulu, Johannesburg, New York, London, Los Angeles,

Paris, Rio de Janeiro, Sydney, Tokyo). The equipment is laid on pallets and may be supplied to any airport within 10 hours, and in the majority of events, the time does not exceed 5 hours.

Figure 13-6   Lifting of aircraft by special air bags
Source: Vepro company

13.8 EMERGENCY SERVICES AND ENVIRONMENT PROTECTION

At the majority of airports, the fire fighting unit is charged even with interventions at ecological accidents or incidents, and sometimes with a preventive activity in connection with environmental protection. An aircraft accident and an intervention by fire fighting units may result in an ecological disaster. Therefore the emergency plan of the airport must address also the issues of environmental protection during the training of the rescue and fire fighting unit and during the intervention.

A long term goal is to substitute harmless or biologically degradable material for extinguishing agents that adversely impact the environment. Substitution of some types of halons which deplete the ozone layer can be made for some complementary extinguishing agents. More than 50 airlines participate in the HAL project, which is resolving the issues of a 100 % recycling of halon gases from fire-extinguishers. The companies Rosenbauer (Austria), Sabo (Italy) and Sicli (Switzerland) are involved in a research project called Eureka for a substitution of the foam forming surface-active agents on the base of fluorine, which impair the environment, by other material, equally efficient but ecologically harmless.

Similarly, during the training of fire fighting units, attention should be paid to possible pollution of the Earth's atmosphere and underground waters. During free combustion of kerosene, thick black smoke is created which may deplete the ozone layer. In the training centre for rescue and fire fighting units at Stansted Airport (UK), aircraft kerosene has been substituted by natural gas which is combusted in a special ring. The training polygon in the Birmingham airport complies with the stringent criteria of environment protection. All the facilities designed for training (an aircraft fuselage, undercarriage and engine, steel wall) are situated in an impermeable basin, covered by a layer of gravel.. Extinguishing agents and the remains of kerosene are drained off from the basin and treated. Kerosene is recycled for the training.

13.9 SUMMARY

Besides the training and equipment of the airport rescue and fire fighting units, the quality of an airport fire emergency service may also be improved by an active attitude towards the passengers and an increase in the fire resistance of the aircraft. The issues of aircraft construction, although very important with regards to the fire emergency service, is not a subject of this book.

Every flight operator is obliged to provide the passengers with a briefing on safety. The majority of them consist of a presentation given by the cabin staff, instructional cards which the passengers have available, or with a video film. It is questionable whether under the conditions of a real accident, a passenger will be able to open the emergency exit on the basis of such a briefing. Many passengers would be uncertain of how to open the door, asking themselves questions like: 'Which direction is it to be opened, inwards or outwards?' 'What strength is to be exerted for turning the handle?', and so on. The crew have to make themselves perfectly familiar with the emergency equipment of an aircraft during their training in emergency procedures. The passengers ought to have a similar opportunity in 'safety centres' established at airports. The emergency equipment of aircraft mostly used by the airport may be installed in those safety centres. The passenger designated to sit next to the emergency exit could 'touch' and try to open it. The question is what attitude the airlines may adopt with regards to the idea of safety centres. Some may think that they could emphasise the fact that accidents happen from time to time, and discourage a percentage of the passengers from using air transport.

14

PASSENGER TERMINALS

Tony Kazda and Bob Caves

Kauffman's airport law:

The distance from the entrance into the terminal building to the airplane is indirectly proportionate to how much time you have to catch the plane.

(A. Bloch: Murphy's Law)

14.1 AIRPORT TERMINAL DESIGN PRINCIPLES

The terminal is often the first point of contact with the country for the arriving passenger. It is a shop window of the country and makes the first and, upon departure, also the last impression on the passenger. From the architectural point of view the terminals have always been and they still are an a show piece representing the best of a particular country. It is, however, necessary to give priority to the functionality of the building by a suitable layout of the terminal. The design of the building depends not only on the number of the checked in passengers but it must also have regard for the type of the airport operation, in particular whether the airport is predominantly an airline hub or is serving mostly local point to point traffic. The basic design characteristics of terminals can be found in the literature (Ashford and Wright, 1992; Horonjeff and McKelvey, 1994). This chapter extends, rather than repeating, the literature by considering specifically the operational implications on design.

The main function of a terminal is to provide a convenient facility for the mode transfer from ground to air transport, and vice-versa. It is also the national frontier for international passengers, so needing to provide all the necessary facilities for this as well as those for the processing required by the airlines. The demarcation between the airside and the landside is usually within the terminal, marked by a security screen, the airside of which is subject to strict control of access prior to the boarding process. In most cases, only passengers are allowed airside, though in the US, where the airlines tend to operate their own terminals, those saying 'goodbye' (wavers) are also normally allowed to go right up to the aircraft gate. The landside is usually open access from where the passengers are dropped off their ground transport, through ticketing and check-in and up to the airside barrier.

On an airport with a great number of transit and transfer passengers the airside transit part of the terminal has to be dimensioned sufficiently and must be equipped with the systems for transportation of the passengers so that it allows them to circulate rapidly and thus ensure short declared times between the connecting flights. The departure and arrival concourses can in this case be smaller. On the airport with mainly origin/destination traffic an increased attention has to be paid to the design of the departure concourse of the terminal.

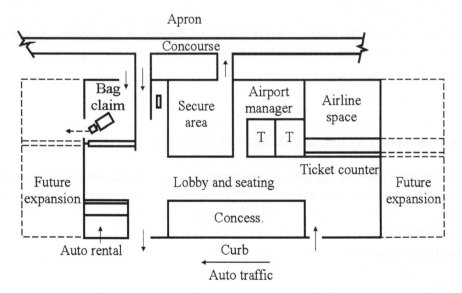

Figure 14-1 **Example of a functional terminal layout**
Source: FAA Advisory Circular, Planning and Design of Airport Terminal Facilities at Nonhub Locations

Passengers and those accompanying them have many tasks to perform within the terminal, the more so with the trend to self-help rather than the use of porters and escorts. They have to find their way through a series of processes, often encumbered by coats, bags, dependants and disabilities. They also have to assist in being processed through check-in, security screening, passport control, the boarding gate, customs, immigration and picking up ground transport, often in a language which is not their first choice. The setting the passengers are given influences them in three major ways: the stress experienced in accomplishing their group or personal goals, the form and nature of their social contacts and their feelings of identity and self-worth. Passengers are most stressed at check-in, security and boarding.

Travellers, airlines and other users of the terminal have their own ideas about the comfort, convenience, costs and ambience that should accompany the movement of passengers and their bags between aircraft and ground transportation. There are inevitable compromises to be made between capital and operating cost, between cost and level of service and between form and function, all of which influence a designer's ability to satisfy all the needs of all users. For passengers to feel at ease in their setting, they need safety, time, the elimination of unknowns, comfort, amenities and beauty. Only when passengers feel safe, know where they are going and have enough time to spare do they pay much attention to the levels of comfort and beauty in the terminal. It is likely that all of the categories will have some effect on well-being, but it is interesting to note that, after the obvious safety requirement is met, the most important categories are time and the elimination of unknowns. Time and its variability are affected by the efficiency of the processing and the speed of moving through the building, the latter in turn depending on the crowding and the ability to find the correct route. The processing and crowding depend on the amount of space provided for a given throughput.

Efficient processing and movement requires good anatomical and physiological ergonomics for the passenger. Some of the common problems are:

Manoeuvring heavy trolleys:

✈ to check-in

✈ along extended 'crocodile' check-in queues

✈ through customs to ground transport.

Check-in:

✈ height of the counter

✈ trolleys moving as passengers try to lift off their heavy bags

↛ impossibility of keeping weight of bags close to passenger's body when placing bag on scales

↛ hearing the operator's questions

↛ service 'justice'– in terms of being processed in order of arrival, which is ensured by 'crocodile' queues at the expense of the appearance of a great queue length.

Movement through the terminal:

↛ slippery floors

↛ lack of assistance with coats and hand baggage – there is a trend to carry more bags on board but few airports provide trolleys in the airside of the terminal.

Despite these problems, the primary factors controlling passenger acceptance are the space provided, the minimum time required and the wayfinding ability.

The design of an airport terminal is affected by the composition of the mix of passengers and by their requirements. In airports with a great proportion of charter flights, the concourse in front of the check-in counters must be sufficiently deep and must offer enough space to cope with the inevitable long queues of passengers arriving in large batches and being processed through a small number of check-in desks. They are normally either migrant workers or package holidaymakers, frequently with over-sized baggage, and often impede free movement in the departure concourse. Airports with a high proportion of business travellers must offer them fast check-in and the shortest possible distance to the planes. It is therefore important, before starting the design of the terminal building, to know the type of airport operation, type of passengers, their composition and their requirements.

However, the type of operation and the passengers characteristics in the airport can change, either gradually or suddenly. Because of these changes, caused not only by economic but also by political factors, the terminal buildings must be designed very flexibly. See Figure 14-1.

An example of such changes might be the introduction of a special channel for checking in passengers of the European Union at the airports of the member states of the European Union or changing concessions layout following the abolition of duty-free sales. Therefore the terminal buildings must provide fast, simple and cheap reconstruction. Changes of the layout of the building must be executed so as not to disturb substantially the check-in or bag claim processes, given the difficulties in changing conveyor routings. From this point of view it is possible to say that, in the past, designs have been too inflexible. Instead of masonry partitions in the terminal building it is more suitable to use light assembled or wood partitions or even 'moving walls', which can be moved within one night or a few hours. It is then necessary to make the service installations in the floor or, even better, in the ceiling of the terminal building. The terminal must be capable of further and simple extension as the number of passengers increase without the necessity of further significant modification of its critical elements such as foundations, staircases and services. The terminal building layout must in particular provide:

➤ terminal – air-side connection

➤ terminal connection to the land-side access system for the surface transport

➤ as short as possible walking distances of passengers upon arrival/departure

➤ information for the passengers during the whole check-in process

➤ transfer and transit of the passengers

➤ appropriate sizing of all areas

➤ required variety and quantity of aeronautical and non-aeronautical services

➤ space for offices and all support services.

14.2 AIRPORT TERMINAL LAYOUT

The terminal building is a connecting link between the air-side the land-side. It is a connection between 'the sky and the earth'. It must provide basic aeronautical services, that is activities directly connected to provision of the passenger check-in process. The building must provide fast and the shortest possible transition of the passengers from the means of surface transport through the check-in process up to the planes upon departure and in the opposite direction upon arrival. The arriving and departing passengers must be physically separated, not only for fast and fluent movement of the passenger flows but also in order to ensure security (see Chapter 15). The flows of passengers can be separated by fixed or moveable obstacles on a single level (horizontally) or on several levels (vertically).

For small airports, a single level concept is the most suitable (Figure 14-2a). In this concept the departing and arriving passengers and their baggage are separated horizontally, mostly on the same level as the apron. The arriving passengers on domestic flights often do not even pass through the terminal building and the baggage is directly handed to them from the baggage cart under a shelter or even near the plane. In this single level concept passenger loading bridges are not used.

If terminal buildings at larger airports were designed on a single level, the terminal would require a large area of land. It is then more convenient to separate the passengers vertically. The simplest type of vertical separation is a concept of one and a half levels in several variants according to local conditions (Figure 14-2b). Normally the departure and arrival landside concourses are on the same level side-by-side. The division of flows of departing and arriving passengers and baggage can be done at any point after the check-in process or alternatively immediately after the entrance into the terminal building. Both levels usually meet again on the apron. This system facilitates the installation of passenger loading bridges, where they are regarded as necessary for passenger safety or as economically viable.

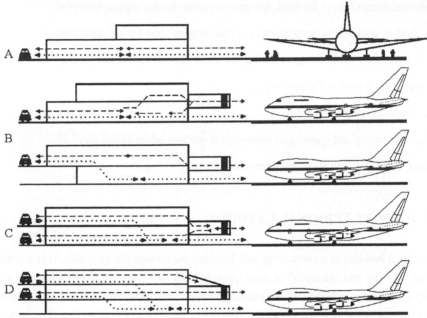

Figure 14-2   Airport terminal layout
a/ single level concept b/ one and half level concept
c/ double level concept d/ three level concept

The double level concept provides separation of the passenger flows even on the landside by vertical stacking of the road access system, though with the capability to move between the levels inside the building (Figure 14-2c). The double level concept is usually used for terminals with traffic volumes of above 5 million passengers a year.

In addition to the flows of arriving and departing passengers the three level concept also separates baggage vertically (Figure 14-2d). It is particularly advantageous to use this concept at airports where the baggage transport system and also other systems for technical handling have been designed below the level of the apron.

The flow of passengers through the departing process must be direct, logical, limiting the changes on the vertical levels and as short as possible. Maximum walking distances for the passengers are recommended by IATA as shown in Figure 14-3 and described below:

✈ from the departure kerbside in front of the terminal building to the check-in counter   20 m

✈ from the farthest car park to the check-in counter   300 m

✈ from the check-in counter to the farthest - gate   330 m

✈ from the gate to the plane   50 m

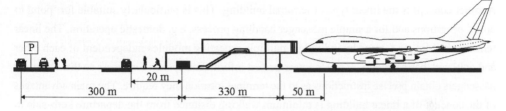

Figure 14-3   Maximum IATA recommended walking distances

The walking distances are similarly specified also in the case of passenger transfer between aircraft. If the distances are longer than specified, it is necessary to provide some kind of mechanical assistance for the passenger (see Section 14.7).

14.3 AIRPORT TERMINAL CONCEPTS

The terminals at small airports have mostly been designed as centralised buildings, that is, where the processing of the passengers is done in one location rather than being distributed through several points in the terminal. The concept of the centralised terminal in combination with piers, fingers or satellites is also used in case of large airports. It provides easy orientation for the passengers, optimum utilisation of space and concentration of services in the terminal building. However, as the number of stands increases, the distance to the outlying stands exceeds the recommended walking distances and therefore it is necessary to provide transportation for the passengers from the central processing building to the gates, together with an effective information system. A central terminal building with a system of several parallel satellite piers interconnected by a transportation system makes an almost ideal solution for large airports. It has a large capacity of both stands and peak hour passengers. It enables transfer of passengers to and from common travel areas without using the central building, which is then not required to handle these passengers. Therefore this design is convenient for the hub and spoke type of operation. At present this design has been used e.g. in the airports at Atlanta, Oslo - Gardermoen and London - Heathrow Terminal 5.

In the past, large airports preferred the concept of several decentralised buildings, with the traffic segregated between terminals by airline, by domestic/international or by some other categorisation of passengers. This is particularly typical for airports in the United States, where the construction of terminals is often financed by airlines. In cases when transfer between buildings is unavoidable, it is often necessary to provide mechanical transportation of the passengers between the terminal buildings in order to minimise the connecting time, e.g. Dallas-Forth Worth airport.

Another concept is the linear type of terminal building. This is particularly suitable for 'point to point' operations and for a simple passenger handling process, e.g. domestic operation. The linear terminal building is formed by individual multiple repeating modules independent of each other and only with basic services. It needs a near-perfect information system for access so that departing passengers obtain precise instructions as to the terminal location they require. The main advantages of the concept of a linear building is minimum walking distance from the departure kerb-side in front of the terminal building into the aeroplane, and simple baggage handling The main disadvantage is relatively high personnel staffing levels, as it is necessary to have a security check in each module, though some airports and governments require a security check at each gate even in centralised terminals. Also, it is practically impossible to provide shopping and catering facilities, which are becoming an increasingly significant source of airport revenues, and it requires dedicated sets of gates for specific flights. This concept hardly still exists in its classical form, except in very small scale terminals, though it can be seen in operation on the landside at Dallas-Fort-Worth. It is best used for domestic point-to-point operations.

14.4 TERMINAL DESIGN

Terminal design can be divided into static and dynamic. Static design is called dimensioning. It enables proportional setting of dimensions for individual parts of the terminal building. The dimensions of individual parts of terminal buildings are usually recommended in relation to the throughput in the peak hour. For dimensioning of terminal buildings some computer programmes can be used, such as IATA Airport Terminal Capacity Analysis/ Airport Terminal Facility Sizing. This allows the estimation of:

✈ the length of the departure kerb

✈ the size of the departure concourse

✈ the number of check-in counters

✈ the number of departure passport control counters

✈ the number of security channels

✈ the size of the departure lounge

✈ the size of gate hold rooms

✈ the number of arrival health control channels

✈ the number of the arrival passport control counters

✈ the size of the baggage claim area

✈ the frontage required on the bag claim devices

✈ the number of arrival customs counters

✈ the size of the arrival concourse

✈ the length of the arrival kerb

✈ the area of the restaurants.

However this model does not include all components of the terminal building, such as amenities, shops, airline offices and other types of catering facilities.

The standard approach to provision of appropriate space is to choose a representative busy hour, accepting that, for the few hours per year for which there will be more traffic, there will be unacceptable levels of crowding, and to follow guidelines for the amount of space per passenger in that busy hour will give a required level of service (LOS). US Federal Aviation Administration (FAA) methods suggest 23 m^2 per Typical Peak Hour Passenger (TPHP) for domestic terminals, and an additional 14 m^2 for international terminals, corresponding to a mid-range LOS of C to D in the International Air Transport Association (IATA) guidelines. The IATA guidelines for space requirements range from 2.7 to 1.0 m^2 per occupant for LOS A to F respectively for long term waiting space, reducing to 1.4 to 0.6 m^2 for gate holding rooms.

These standard design procedures for airport terminals are not at all satisfactory. They are based on handbook formulas insensitive to the realities of each situation, and the formulas are easy to misunderstand and thus frequently misapplied e.g. the common design hour LOS standards should apply to simultaneous occupants (except the FAA guidelines per TPHP), but are often used per design hour passenger even if their dwell time is much less than an hour. They should be improved by a realistic appreciation of the dynamics and behaviour of sequences of queues, the psychology of crowds in such situations, and the ways airport users truly allocate the time they spend in passenger terminals – their slack time is often much greater than their processing time. People do not spread out evenly like gasses, but tend to congregate in specific places, e.g. the mouth of the baggage chute in the baggage reclaim hall. LOS standards should increase with exposure time. The IATA guidelines for space for adequate LOS C appear, in many cases, to be over twice that necessary, apparently being defined for the case where almost all the passengers are in movement and have baggage trolleys. The requirement is also strongly influenced by whether the activity is to be joint use or dedicated to an airline, a route or a type of operation. Territoriality in public spaces leads to a wasteful use of public facilities.

'Intensity of use' and 'service time' are less value-laden terms than 'crowding' and 'time delay', and may therefore be more appropriate as general measures of performance. Because passengers are often accompanied by visitors, total numbers of people accommodated may be a more appropriate measure than passengers per peak period. Current measures of performance based in the space requirements for a standard 'busy period' are poorly suited to terminals serving intensive hub-and-spoke flight schedules, aircraft fleets with large fractions of high capacity wide-bodied and commuter aircraft, and growing numbers of less sophisticated passengers. In fact, there is a large variety in passenger perceptions of adequate space. The length to which an individual will go in accepting cheerfully what most would consider to be extreme inconvenience is remarkable, so long as it is his/her own choice. This is reflected in the recent calls by the low cost carriers for the minimum of facilities in new terminals in order to keep charges down.

The accuracy of the model results depends also on the availability and suitability of the input data. If there is insufficient data there can be only tentative results; for example, in the design of the kerb length in front of the terminal building it is necessary to know the split of the passengers arriving to the airport by car, taxi or by bus, the average number of the passengers in each of the means of transport, average time of stay for each vehicle and the length of waiting space necessary for each of the means of transport. However, the length of the kerb can also be designed using the default values for all kinds of vehicles. For small airports, under 0.5 million passengers per annum (mppa), it is advisable to compare 'rules of thumb' values with expert judgement.

Some computer models are able not only to size the individual parts of the terminal building, but also to consider the location of the facilities in the building. For example, to maximise profit levels, shops should be located so as to be visible as the departing passenger moves through the building on a natural path towards the aircraft. Passengers are quite likely to buy on impulse, though it has to be said that, under 1 or 2 mppa, it is difficult to raise much revenue from terminal concessions.

For airports with passenger volumes from 0.5 to 5 mppa, the use of static, or fixed ratio, models is problematic. The volumes of the passengers are already too big to assess the data obtained form static modelling by expert judgement, but too small for the system of the airport to be relatively stable. Therefore it is more appropriate to use simulation models enabling dynamic modelling for such airports.

Dynamic design, simulating individual parts of the terminal building, allows individual parts of the building to be considered operating under different conditions. Thus, for check-in counters, it is possible to set different lengths of the queues for the different categories of the passengers (3 passengers in the queue for the passengers of the first class, 5 passengers in the queue for the business class and for 90 % of all passengers a waiting time less than 15 minutes, etc.)

14.5 THE HANDLING PROCESS

14.5.1 Passenger Handling

The concept of the departure or arrival handling processes can be different in several ways. The most basic difference is the handling of domestic and international passengers due to differing border controls. There are three kinds of borders in Europe at present:

✈ borders between the countries of the European Union and the third countries - they have the same character as in the past

✈ borders between pairs of countries of the European Union - the movement of goods is free, the movement of persons is inspected

✈ borders between the countries within the European Union that have signed the Schengen Agreement - these are treated as domestic movements.

Other differences result from the customs, passport, safety and health regulations for individual countries, like the special security requirements for Irish routes to and from the UK. Upon departure two basic passenger check-in concepts are possible: **check-in for individual flights** and **common check-in.**

With the individual flight check-in concept the passenger for a particular flight can be handled at one or several counters reserved for that flight. The data on the passengers and on the flight booking status, which are necessary for the bag loading, can be collected manually or taken from the reservation system. Often the advantage of this procedure is the shorter handling time for an individual passenger, because the check-in of an individual passenger does not require access into the central booking system. Charter flights particularly tend to use this type of check-in, but with few desks. The large number of bags can result in long queues.

In the case of the common check-in concept, the passenger can check-in at any counter of the given airline, handling agent or even in the whole terminal building or that part which is suitably equipped. It requires computer technology to support this method of handling. Information on the passenger (flight booking status) must be available at each check-in counter. Computer technology enables also further data processing in the course of the check-in process (seat allocation, data for aircraft loading, catering special requirements, etc.). The advantage of the common check-in is equal load on all check-in counters. In some cases the disadvantage is relatively longer handling time of the passenger flights which would have had short queues and, if the handling agent provides the check-in, there may be a loss of identification of the passenger with the airline. The same counter is often used during the day by several airlines or handling agents. This is made possible by the CUTE system (Common Use Terminal Equipment), which provides access into the computer network and makes it possible for each company to use special software. The logo of the company can be displayed above the counter during the time it is actively using it. It may also be used by a single company with many flights, in which case the queue may take the form of a crocodile, particularly where the available depth for queuing is small. Passenger avoid being held up in a slow queue, but have to move their luggage forward many times.

The passenger handling process on departure is normally for the passengers to submit their air tickets at the check-in counter and their passports for control. This allows their booking status and identities to be checked. The checked baggage (the baggage transported in the aircraft cargo holds) will be weighed and, if the free weight limit has been exceeded, the appropriate fee will be charged. The weight of the baggage will be registered into the air ticket and baggage receipts stuck on the air ticket. The passengers will get their boarding passes. On some flights the airlines use a piece concept. In this case passengers are allowed to take one or more pieces of baggage of specified dimensions free of charge. All baggage consigned to the hold should undergo security check (e.g. X-ray). Within a few years 100 % security hold baggage screening will be compulsory in most states. The security check of registered baggage can be made before the check-in or in directly at the check-in so that the passenger could participate in it. However, automated security checks can be done at any part of the baggage handling process. Some countries require also outbound customs inspection. Passengers then move through to the emigration passport control and security checks of the passenger and any cabin or carry-on baggage.

In order to speed up the passenger handling process and decrease the requirements on the number of employees, an Automatic Ticket and Boarding (ATB) system has been introduced for passengers with no hold baggage on some airports. By means of a touch screen the passengers can find out, whether there are free seats on the particular flight. The ATB machine will print out air tickets for the passengers, for which they will pay by credit card, together with boarding passes with magnetic strips for entry to air side. ATB provides positive identification of the passengers and their baggage.

The sequence of individual stages of the check-in process can be different at various airports. In some airports the central security check has been placed at the entrance into the check-in area and only those holding an air ticket are eligible to enter the check-in area. Many other combinations are also possible, including allowing check-in at the departure gate.

The system of check-in immediately after getting out of the car on the walkway in front of the terminal building, **curb check-in,** is common in the airports of North America. In this way, passengers are relieved of their baggage and may complete the check-in at the counter in the building.

Many of the new low cost carriers are now not assigning seats, so, for domestic flights, the only reason for checking in is to consign bags to the hold.

On arrival, the passengers first go through the passport control. The passengers can be divided into several channels, according to their nationality and visa regulations. For business or frequent travellers, the system of automated control is being introduced in some airports. Upon the first entry into the country and upon request, passengers are issued with an identification card, which, in addition to the basic passport data, contains also the description of biometrics (e.g. dactyloscopic marks of the hand, face geometrics or iris characteristics). On later entries a reading facility checks the biometrics of the passenger with the data on the card. A health control follows the passport control if necessary. After the baggage has been collected by the passengers on international flights they go through the customs. In order to speed up the customs inspection a concept of red and green exit is commonly used. The red exit is used by the passengers, who declare the goods at the customs'. The green one is used by the others, only spot checks being made. A blue exit can also be found at the EU airports. The EU nationals, to whom special customs regulations apply, use it.

It is, of course, necessary to find one's way between these processing points. Perfect wayfinding will result in a minimum necessary time for the journey. Any lack of wayfinding ability will increase the uncertainty in predicting the needed time, as well as increasing the average time taken. This is particularly important because a prime psychological concern is to eliminate the unknowns. Some of the uncertainty comes from the difficulty in estimating the effect of the possible barriers to processing and movement, and some from the difficulty in actually navigating though the terminal. It is recognised that the spatial logic of a natural line-of-sight progression in a straight path from ground to the aircraft and vice-versa is the ideal design. This is easy to provide when traffic levels are low, or if a decentralised 'gate-arrival' design is used to minimise walking distances. It is, however, very difficult to provide natural wayfinding in terminals for millions of passengers per year without the depth of terminal becoming prohibitive. This is the concept used at Stansted for eight million passengers per annum, but it is imperfect because of all the ancillary activities which have to be fitted in on the single floor level and also because the view of the aircraft is illusory in the sense that they are not accessible directly from the terminal but only from the remote satellite terminals via a people-mover. It has also been used in principle for the new Hong Kong airport at Chep Lap Kok for 30 million passengers per annum. The eye is, in fact, drawn to the airside end of the pier by a descending roof rather than a direct view of the aircraft. However, the forward motion is impeded by having to cross long check-in queues which form at right angles to the flow from kerb to aircraft, and by the need to change levels to access the people-mover.

When complex routes through a terminal become necessary, an efficient signing system becomes essential. Considerable effort has been put into understanding the best way of communicating information about flights to passengers, and into enumerating the advantages of TV screens or large flip board displays relative to verbal announcements. IATA and ICAO have set recommendations for symbols to try to cope with the language problem, and most major airport groups have their own brand of shape, colour and font for written directional signs. However, it is difficult to avoid the impression that all these signs have been developed by air transport personnel without consulting the users. The words 'departures' and 'arrivals' do not mean the same for everyone, nor is an upward-angled arrow or aircraft symbol always interpreted as a need to change level. There are also the perpetual problems of avoiding clutter with too many signs, differentiating between wayfinding and other signs, preserving visibility in crowded conditions with low ceilings, and assisting passengers who need to back-track. The new terminal at Vancouver uses lighting graduated in brightness towards the nodes for subliminal wayfinding, and also uses the light fittings and carpet markings used as pointers

Wayfinding and being processed are both more difficult for the disabled. New legislation is going to require airports to differentiate as little as possible between the disabled and the able-bodied. The ICAO definition of a person with disabilities is 'any person whose mobility is reduced due to a physical incapacity (locomotory and/or sensory), an intellectual deficiency, age, illness or any other cause of disability when using transport, and whose situation needs special attention and the adaptation to the person's needs of the services made available to all passengers'. Up to 20 per cent of the first world population has a significant disability, plus many more who do not report their disabilities. Each type and grade of disability has its own set of ergonomic challenges. It is an airline responsibility to provide the necessary facilities from arrival at the airport to boarding the aircraft, for example, scissors lifts for separate and early boarding, the service often being subcontracted to specialist suppliers.

If it is at all possible, respect for the dignity of the disabled should be preserved, and this is best done by making it possible for them to use the same facilities and routes as other passengers by providing the same information and accessibility. Considerable progress has already been made in making physical provision by ramps, lifts and minimising changes of level, and now with aids to wayfinding. Some recent helpful initiatives are:

For the blind:

✈ synthesised voice calling out the floor and direction in all elevators

✈ tactile maps of the terminal.

For the visually impaired:

✈ high contrast flight information displays

✈ high contrast wayfinding information embedded in the flooring.

For the deaf:

✈ visual paging systems displaying text versions of audible announcements

✈ TDD service on public pay phones and at airport customer service points.

For the hard of hearing:

✈ amplified handsets on all counters

For all requiring carers:

✈ unisex washrooms in every cluster of washrooms.

Despite this progress, much more needs to be done if the disabled are really to be able to be as independent as the fully mobile.

14.5.2 Baggage Handling

Baggage handling is becoming a critical activity. The airlines are trying to shorten the turn round time between individual flights and at the same time, the average load factors are increasing. On big airports there may be on average 210 passengers per aircraft. If the free baggage allowance of 20 kg of hold baggage per passenger is taken up, in the course of 40 minutes approximately 8 tonnes of baggage must be checked in, weighed, loaded and unloaded. One of the criteria for quality of service of the airport is the time passengers have to wait for baggage after disembarking from the aircraft.

At small airports most of the activities connected with baggage handling can be carried out manually. The baggage is loaded on the baggage cart either directly or from the belt conveyor and dispatched to the aircraft, where they are loaded in the aircraft cargo holds. The baggage can even be handed directly from the baggage carts to the passengers. On medium size airports with the departure of several flights in the same time, sorting of baggage must be provided. Sorting can be done manually, when the baggage is picked up from the carrousel and loaded on the carts, or semi-automatic, in which case the operator will pre-select an appropriate branch, to which the baggage will be dispatched, according to the baggage tag. In both systems, because of the human factor mishandling can occur, leading to a certain number of lost or mislaid bags, which damages the reputation of the airport.

Figure 14-4　At hub airports passengers require wide range of non-aeronautical services, photo A. Kazda

In big airports it is necessary to provide automated sorting of baggage. Automation can increase the capacity of a terminal building significantly and also improve the service standard. Sorting uses baggage tags with bar codes, magnetic cards or electronic chips for baggage and destination identification. Bar coding is the most frequently used today. In this case the tags are automatically printed when passengers check-in. Reading of the code is carried out automatically, by laser sensors, alternatively semi-automatically with manual laser sensor and the data are stored in the database. The data are used for the baggage routing within the baggage sorting system. These data are also compared with the data from the passengers database. This provides positive identification, called baggage reconciliation, to ensure that passengers have boarded the same aircraft into which their baggage has been loaded. If any passenger did not board the aircraft it is possible to determine exactly where the baggage is and unload it for security. On airports with a great number of transfer passengers it is moreover necessary to provide sorting and redistribution of baggage in a very short time in order to guarantee the minimum transfer times of the passengers. It may be best to provide sorting of transfer baggage outside the main terminal building, particularly for online transfers within an airline's hub.

14.6 Non-Aeronautical Services

If we consider the needs of the passengers, it is necessary to realise that the main reason why most people come to the airport is not the airport itself, but because they want to get somewhere else. Most of the passengers are in a condition of emotive stress, even when they have already travelled by plane before. Therefore it is necessary to create a familiar and pleasant environment for the passenger, where he or she could relax and spend time before boarding. The airlines will provide this for premium passengers in their own lounges, but airports have seen the need to provide for all passenger types, and the opportunities that this brings.

At present many airport administrations achieve higher revenue from other than the aeronautical services. This is caused by several factors. One of the main factors is the need to increase revenue. The possibility of revenue increase from landing and handling fees, is in many countries strictly limited by the regulations of antimonopoly authorities and also by the strong lobbying of air carriers represented by IATA. At the same time the advantage of providing non-aeronautical services at the airports is that on the airports there is a high concentration of the more affluent members of society, and none of them is able to completely avoid the dwell time in the terminal before boarding the plane. Therefore during this time he or she can use shops and services, if they are some available and interesting, thus raising revenue for the airport and the concessionaires (the shop operators).

The minimum provision of aeronautical services is explicitly determined by technological requirements and by the scope of the airport operation. The extent of non-aeronautical activities depends more on the space available at the airport and on the passenger volumes. Sometimes non-aeronautical services can reach such a level that they impede the handling process. For example, they can limit the capacity of check-in counters, interrupt the flow of departing passengers or their view of the location of toilets or departure gates. In spite of the fact that diversification and variety of offered services is required, non-aeronautical activities should never disturb smooth running and basic functions of the airport.

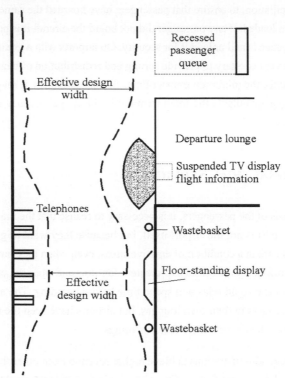

Figure 14-5   Public corridor effective width

It is, however, possible to take advantage of the increased scope of services to make the airport operation easier. They can divert the passengers from the main circulation streams in the airport and help to distract the attention of the passengers from the problems and irregularities in the air transport system.

On some airports the traditional lay-out of the check-in process and non-aeronautical services is changing. Figure 14-6 shows the traditional lay-out and a new approach to placing of services in the terminal building:

In this way, the passenger, passing through the terminal, actually goes through the shops. Some say: 'This is not an airport any more, this is a shopping mall' however others claim: 'Only now it is beginning to be an airport'.

The needs and wants of the passengers gradually change. It is certain that typical passengers today expect to be offered more than last minute duty-free shopping at the gate. The passengers will require a wider range of services: hotels business services, various catering possibilities of the same standard, quality and price to which they are accustomed. Already airports have provided additional services, from fitness centres to super markets, where the customers can choose from the same range of goods as in the high street shops with brand wares and fashion goods.

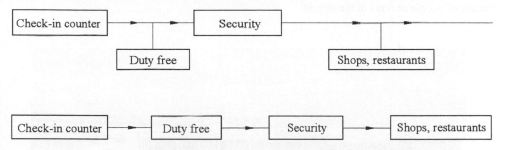

Figure 14-6   Change in the lay-out of the check-in process

Though the airports' characters gradually change and become similar to the city centres, it is apparent that they will always differ by the scope of offered services, depending on the specific role within the catchment area (transit airport, origin/destination or hub airport).

14.7 PASSENGER TRANSPORTATION - PEOPLE MOVERS

The use of widebody jets at the end of the sixties necessitated extension of airports, particularly of terminal buildings. Greater wing spans required greater distances between the stands and longer walking distances for the passengers. On the other hand, economic pressures have forced the airlines to increase the daily utilisation of aircraft and to shorten their turnround times. At big airports this necessitated the introduction of passenger transportation. Most of the means for transportation of passengers, which are used in the airports, were originally developed for transportation in cities. Some of them were improved or modified for use in the airports.

As mentioned above, there are internationally recommended maximum walking distances for the airport buildings. Usually this distance is not supposed to exceed 300 m. If it is greater, it is necessary to provide assistance for the passengers. The second task of transportation facilities for the passengers in the airports is to minimise the connecting times between individual flights. The generic term used to describe all facilities used for passenger transportation in the airports is people movers.

Terminal buildings on the airports with a volume of up to 5 million passengers a year generally do not require installation of people movers. Though it might be desirable to facilitate movement of passengers on the airport, each people mover means an increase of investment costs in the construction of the airport. It means also an increase in the operational costs, particularly to guarantee the availability and reliability of a facility. Also, passengers who use a people mover cannot use the shops. Therefore, if it is acceptable for the passengers, there is a tendency to avoid the use of people movers in the airport.

Figure 14-7   Mobile lounge, photo A. Kazda

The choice of a suitable people mover system depends on:

✈ speed

✈ capacity

✈ safety and security.

Other criteria for the choice are usually these factors:

✈ transportation distance and elevation difference

✈ required frequency

✈ reliability

✈ ease of use by the handicapped

✈ ease of use with accompanied baggage

✈ maintenance requirements

✈ design characteristics

✈ procurement and operational costs.

Figure 14-8   People mover system connecting airport satellites
photo: courtesy Orlando Int. Airport

For transportation of the passengers between the terminal building and remote stands on the apron or between individual buildings on the airport it is possible to use regular buses or special buses. The special airport buses used for the operation on the apron usually have a bigger capacity than regular buses. Because they have not been designed for regular operation on public services, they can be wider and have lower clearance. This makes it easier for the passengers to get out and to get in. In spite of the fact that special buses have higher capacity, sometimes the capacity of one bus is not sufficient. Therefore in Honolulu airport the system of 'bus train', is used, which consists of several units with a drive and semi-trailers. For the transportation of the passengers from the terminal building to the plane some airports use mobile lounges. The advantage of the mobile lounge is improvement in the utilisation of the apron and simplification of the passengers movements. The passengers do not have to change level as when using buses. If all the stands on the apron have been designed as remote and mobile lounges are used for transportation to them, there are advantages of a quieter and less polluted environment in the terminal building. The disadvantages are the higher price of mobile lounges in comparison to the buses, and the total time of transportation in the mobile lounge usually is, in comparison to the buses, longer (passengers get in and get out only through the door, which is placed in the front part of the lounge). All these special vehicles have implications for the design of the airside roads. In fact, whatever means of surface transport is used to access remote stands, there are serious implications for space, safety, pollution and minimum connecting times.

The simplest and also the most widely used types of people movers in the terminal buildings are escalators for overcoming changes in level and moving walkways for near-horizontal transport. Moving walkways are mostly used for distances up to 200 m. They usually do not significantly shorten the time of the passengers to reach the aircraft. The usable distance is limited by the walkway speed, which usually must not exceed 1.25 m.s^{-1}. The length of walkway is also limited by the fact that it is only possible to get off of the walkway at its ends. Therefore several sections of walkways following each other have to be installed in the corridors to the gates. This is an incidental benefit when one section is out of action, in that the quality of service is not too seriously affected.

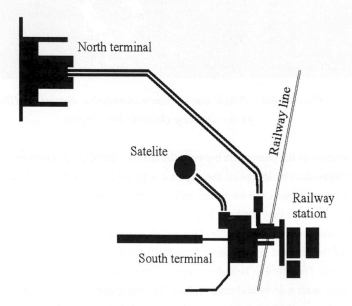

Figure 14-9   Scheme of people movers at Gatwick airport

For transportation of the passengers between individual buildings over longer distances at the airport, e.g. between terminal buildings (Gatwick, Kuala Lumpur) or between the terminal building and the railway station (Birmingham) the use of walkways is not appropriate, because of their slow speed. In these cases it is normal to use an automated shuttle type of people mover either on one route or on parallel routes.

The first shuttle type people mover in Europe was installed at Gatwick airport. The part connecting the new satellite with the existing main terminal building and the railway station was built first. Later another route was built, which connected the south terminal with the new north terminal Figure 14-9).

A people mover with the highest passengers throughput is the system at Atlanta Hartsfield International, which transports on average 109 000 passengers a day (Figure 14-10). The system connects the four parallel satellite piers with the main terminal building. The people mover is located in a 1 600 m long underground tunnel together with the baggage transportation system and a walkway.

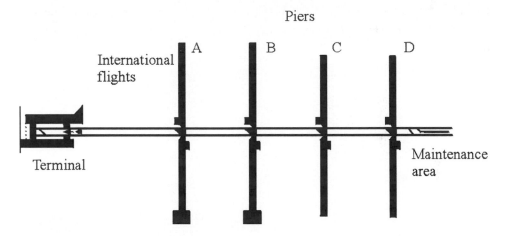

Figure 14-10 People mover system at Atlanta Harsfield airport

The reliability of these types of people movers is usually higher than escalators or movable walkways. In the transportation peak the people-mover operates at set intervals. Outside the peak it may be possible to call the vehicle by a pushbutton as with an elevator. There are many different types of constructions and drives. The Westinghouse company uses fully automated carriages on tyres with electrical drive, while the OTIS company uses its own technology, which is used for elevators, the carriages being driven by a steel rope.

On large airports, providing surface transportation of the passengers is just as important as the air transport. Often the surface transportation of passengers at the airport is the bottleneck, which limits the airport capacity. This is considered further in Chapter 16.

15

SECURITY

Tony Kazda and Bob Caves

15.1 UNLAWFUL ACTS AND AIR TRANSPORT

The expression 'security' concerns all unlawful acts connected with civil air transport. The character and danger of unlawful acts change gradually not only with the change of exterior, particularly political conditions, but the terrorists react also to the adopted measures for safeguarding the security of air transport.

The level of safeguarding security is one of the important factors by which the airlines judge an airport. An aircraft of the American airline TWA was hijacked in Athens in June 1985. During the high-jacking one American national perished. Five months later the terrorists hijacked an aircraft of the airline Egypt Air to Malta. In order to achieve their aims, they started systematically to murder the passengers at regular time intervals. To rescue those who remained alive, a commando-style rescue operation was carried out. The result was 61 dead and 21 injured. The aircraft was completely burned out. After this event, American citizens were warned not to fly to Athens and Greece, since it was not capable of ensuring their security. The number of tourists from the USA in Greece decreased by half during the following tourist season.

With unlawful acts, the actual number of casualties is less significant than the threat that everybody who uses air transport could become this casualty. For example, a total of 30 citizens of the USA were killed in the terrorist acts in 1985. Is it possible to compare this number with 100 casualties killed by lightning, with 1 800 murdered and approximately 40 000 killed in road accidents? It is the violence and insanity of terrorist acts that is so disturbing, together with the fact that the victims do not get the chance to defend themselves against these acts.

It is necessary to realise that air transport is not the real target of terrorists. The targets are 'enemy' countries and their governments, upon which the terrorists want to enforce a change in their politics. To attack the 'enemy' country, the terrorist does not have to risk performing sabotage activities on the territory of a foreign country. On the contrary, he can wait that the 'country', represented by an aircraft of its carrier, comes to meet him. The terrorist himself can choose the time and place for the attack. The aircraft itself may be worth several hundred million dollars. The unlawful act will become a central theme of the news of all television and radio stations of the world for several days, and in this way the terrorist will get the required publicity.

The first recorded unlawful act in the history of aviation was hijacking of an aircraft in Peru in 1930. The hijackers used the plane to drop political leaflets. The first casualty of an unlawful act was the pilot of a Romanian aircraft which was hijacked in July 1947 during a flight from Romania to Turkey. The third case happened in April 1948. Seventeen hijackers from Czechoslovakia, among them were also two crew members of the plane, forced the plane to land in the American occupied zone of Germany.

Hijacking of aircraft was practically the only type of unlawful act until 1969. The hijacker needed the plane as a 'means of transport'. The highest number of hijacks was from the USA to Cuba. It is evident from the analysis of hijackings that there was a close dependence between the state policy limiting the free movement of citizens and the number of hijacks. After opening the frontiers the probability of 'traditional' hijacking decreased.

Where the politics cannot be changed, measures have to be introduced to counter the terrorists. The number of hijacks in the USA decreased substantially after the introduction of security measures. The measures included manual inspections and checks of passengers by metal detectors and inspection of their cabin baggage. The checks were supposed to prevent the would-be hijackers from carrying weapons onto the aircraft.

After 1968 the character of unlawful acts started to change. Terrorists began to use hijacking of planes for enforcing their demands rather than for transport. Hijacking was a suitable tactic for terrorists because:

✈ the passengers and crew were suitable hostages

✈ the aircraft was a suitable, safe and temporary mobile prison for the hostages

✈ through the media that monitored the case it was possible to lay demands upon authorities.

The countries in the Middle East and in Europe were not able to react quickly and effectively to hijacking and they were slow to recognise the problem represented by transit and transfer passengers.

Terrorist tactics changed after 1986. Increased security measures made it more difficult to carry the weapons onto aircraft. Therefore the terrorists adopted methods which were easier and safer for them. Before 1986 sabotage as a coercive measure was seldom used. The difference from hijacking was that sabotage did not make it possible to keep hostages, to put conditions and to blackmail authorities. Those sympathising with the terrorists realised that they could also become victims. The responsibility for the committed crime could also be claimed by other 'rival' terrorist groups. In this period, sabotage, as distinct from hijacking, was considered to be a condemnable act in the countries where the terrorists came from.

From the beginning of the 1980s the number of unlawful acts decreased but the consequences of the terrorist actions became more severe. The formation of fundamentalist groups supported by extremist countries changed the situation. Terrorism became an integral part of political and ideological struggle. Terrorists began to use sabotage and sabotage threats against aircraft and airports as coercive measures. Aircraft were destroyed in the air and on the ground. In Jordan at Dawson's Field three aircraft were destroyed on the ground; in Srí Lanka a bomb was placed in the aircraft tail during refuelling; there were massacres in the check-in concourse at the airports of Rome and Vienna in the period before Christmas; there were explosions of an Air India 747 off Cork in Ireland, a Korean Airlines 747 above the Andaman Sea, a Pan Am 747 over Lockerbie in the UK and a UTA aircraft in the desert between Chad and Niger.

The original measures used against hijackers proved to be insufficient against terrorists. The terrorists used more sophisticated means including highly effective plastic explosives, firearms made of plastics and composites and masked bombs. The measures for ensuring security were therefore complemented by inspection of all registered and hand baggage, air cargo, galley equipment, mail etc. Airport employees were included in security inspections and also all other persons that entered the air side area of the airport. Computer controlled systems were used to an increased extent for inspection of authorisation of entry into specific areas. Also hitherto accessible parts of terminals, workshops and maintenance areas on the airport were hereafter included in the security cordon.

A different threat is mortar attacks on airports from beyond the borders of the exterior perimeter of the airport. The first mortar attack on an airport was launched in January 1975 on Paris - Orly airport. After the Irish Republican Army mortar attacks on London Heathrow airport in March 1994, it seems that increased attention must be also paid to the land around the airport, particularly to the car parks or unattended areas.

The advantage of the terrorists is that they themselves choose the manner and time of fighting. It is difficult to presume what the future weapons of terrorists will look like. State sponsored terrorists have access to technically sophisticated weapons and they go through professional training, which enables them to overcome more and more sophisticated counter measures. It is possible that, after further tightening of security measures in the airports, the terrorists will be e.g. able to use controlled earth-air missiles fired from the arm, either Russian SA-7 or SA-14, or missiles from the USA production such as Redeye, Stinger RMP/POST or French Mistral, all with a range of around 6.5 km and a ceiling of 6 000 m. This means that an aircraft could be hit while still far from the airport.

As has been mentioned, the purpose of most of the attacks against civil aviation is to create political pressure on individual governments. The government politics directly conditions the level of threat to civil aviation and a single unwary declaration of a high government representative can mean an immediate threat of terrorist attack with extensive negative economic consequences for airports and civil aviation in general. The protection of civil aviation must therefore be an integral part of national security of each country. It is incorrect that the protection against unlawful acts be paid by airports, airlines or passengers, who themselves are jeopardised as a consequence of the political attitude of the country. Moreover, under international law the governments must ensure protection for all companies and individuals that are on its sovereign territory without any discrimination. In spite of this, most costs in connection with ensuring the security of civil aviation are born either by airport administrations or by airlines. In the long run, the costs are transferred onto the passengers. Some airports have a directly specified security charge.

The threat of unlawful acts is at the same time a very effective way of waging war against another country. It is estimated that the total damages to civil aviation of the USA in connection with the war in Persian Golf exceeded the costs spent on military actions.

The method of ensuring the security of civil aviation is different from country to country. For example in the Federal Republic of Germany the responsibility for security of airports is under law divided among three parties:

+ the government is in charge of inspection of passengers and their cabin baggage

→ the airport is in charge of airport areas behind the security check and the whole premises of the airport

→ airlines assure compatibility of individual registered baggage with passengers and security of their own ground equipment.

International cooperation and standardisation is important as well as coordination of procedures between individual countries in order to ensure the security of civil aviation. The issues of security were for the first time addressed at the international level in the agreement in Tokyo in 1963, which deals with the issues of acts 'disturbing the security of aircraft or travellers'. The growth of politically motivated terrorism in the sixties and unlawful acts were addressed by the Hague Agreement of 1970. The Montreal agreement of 1971 dealt with acts of sabotage. ICAO standards for security issues are dealt with in Annex 17. Annex 17 was published for the first time in August 1974. Since then it has been amended when necessary.

15.2 THE AIRPORT SYSTEM AND ITS SECURITY

Fast changing security threats require different design and construction of terminal buildings and other areas of the airport. The earlier terminals only seldom comply with today's requirements. Old terminal buildings mostly do not allow separation of departing and arriving passengers, so that temporary solutions have to be adopted in order to separate the flows of passengers. It is also problematic to estimate which security requirements will have to be met in the future. It is therefore necessary to design the terminal buildings with the maximum flexibility.

Measures to combat unlawful acts required substantial intervention into the check-in process and into the design of the airport terminal, which originally had not been taken into account. The first measures consisted in setting up security control of passengers and their cabin baggage in the gate, immediately before boarding the plane, or alternatively directly before boarding the plane upon the exit from the gate. For flights with special security requirements, the inspection of registered baggage was carried out bag by bag. It was possible to retain the air cargo in the warehouses for some time. However, these measures turned out to be impractical. When the inspection of the passengers in the gates took place, there was often delay to flights. Decentralised check locations had to be established at the entrance to the holding area for each gate. This increased demands on personnel at the check points and on their technical equipment requirements. The queues of passengers before the security control prevented other passengers from moving freely.

Therefore the centralised system of security inspections tends to be given preference in the design of new terminal buildings. Thus, by the time the passengers reach the entrance to the gate (alternatively to the transit concourse) they will already have passed through one or several security filters. The centralised system has several advantages:

→ passengers inspection is carried out before the entrance into the airside circulation area, so the passengers do not have to wait until their flights have been announced

→ there is a low probability of the flight delay as a result of security inspection of the passengers

→ the flow of passengers through the security inspection is substantially more stable with smaller peaks than at individual gates

→ there is higher utilisation of technical equipment and personnel.

The centralised security inspections have also their disadvantages:

→ corridors or concourses to the gates must be kept security-sterile

→ the security-sterile area must be equipped with the required services

→ before entering the sterile part all employees and goods also have to be checked

→ people accompanying the passengers are prevented from having free access to the whole terminal building; this decreases the commercial use in that part of the terminal. Exceptions need to be made for elderly and handicapped passengers, who, at least in the USA, are accompanied by the dependants up to the entrance to the aircraft.

In some airports, where it is necessary to introduce a special security mode, the preliminary security inspection of all persons is carried out immediately on their entrance into the check-in hall.

Security requirements are often contrary to the architectonic intentions of the design of the terminal building. Balconies, terraces and entresols, which divide the internal area of the building in a suitable way, could be convenient observation points for the terrorists, or a place from where shooting could take place. In the event of a bomb attack, large glass areas, providing natural light, can be very dangerous. Glass shards are a source of extensive injuries. In the check-in hall and in other passenger areas, controls should ensure that it is difficult to leave baggage which might contain a bomb.

Some older airport terminals have been combined with car parks (Toronto, Charles de Gaulle - Aérogare 1). If a car were to be placed there, and the explosives it contained were detonated, there would be extensive casualties and damage. In a period of increased danger of terrorist attacks, the floors of the garages directly adjacent to the areas for the passengers must be closed to the users and modified so as to limit the spread of the detonation wave. In the design of new airport premises, the parking places should principally be located with no direct contact to the terminal building, whether they have been designed as parking places in the open or in multi-storey garages. However, it is necessary to find a compromise, so that the walking distance does not exceed the recommended limits. At the same time it is necessary to provide inspection of parking places, not only to provide security, but also as a prevention against theft.

It might well be advantageous for a terrorist to place an explosive in the left-luggage rooms of a terminal building. Therefore, the baggage must be checked before accepting it into custody. If the airport uses lockers for baggage, they should be placed outside the terminal building.

For flights with special security requirements, it is appropriate to reserve one part of the check-in counters with a separated entrance for the given flight. The counters must be physically and visually separated from other parts of the building. The formation of a queue in front of the entrance to the counters must be eliminated, so that no attack could be made against it. In some airports, a whole terminal has been assigned to flights with special security requirements. This gives the possibility to provide increasingly complex security standards, but there is a clear demarcation of the premises which makes the target obvious to potential terrorist.

It is necessary to pay special attention to maintaining security during construction or reconstruction of the airport premises. The workers of construction companies are sometimes not too willing to undergo the security measures. It is very difficult to find out whether they are criminally unimpeachable. Because of a generally high turnover of staff, it may not be difficult for a terrorist organisation to infiltrate their members into a construction company. Inspection of material movement is made difficult by a great number of cars and delivery vehicles. The only solution is a consistent separation of the areas where the work takes place from other premises of the airport. Even then, it is not possible to exclude the possibility that weapons or explosives might be smuggled into the airport and placed there for later use.

An increased attention to maintain security must also be paid during emergency situations in the airport (fire in the terminal building, emergency landing of an aircraft, etc.). Emergencies can be evoked intentionally in order to distract the attention of security units from a terrorist attack.

15.3 SAFEGUARDING OF AIRPORT SECURITY

15.3.1 Security as a Service

In civil aviation, safeguarding security has to be considered as a service to passengers. A feeling of safety is one of the basic needs that must be satisfied. However what one category of passengers considers to be adequate safeguarding of security, e.g. families with children, can be considered as a useless nuisance by other groups, e.g. business passengers. Therefore a compromise has to be searched for, between the level of safeguarding security, the time required for security inspections of the passengers, and the money costs. The required security standard has to be safeguarded in the airport. On the other hand the airport operation must not be paralysed, nor must the high quality of service of air transport be affected by disproportionate prolongation of the check-in process in connection with introduction of strict and time consuming security inspections.

The system of safeguarding security is individual in each airport. It will be different on large international airports such as London Heathrow, New York J.F. Kennedy or Amsterdam Schiphol from small airports serving perhaps two or three holiday charter flights daily. However the requirement has always to be the same, namely to prevent the possibility of locating a bomb on board the aircraft or to prevent the airport premises being penetrated or aircraft being boarded by a group of terrorists.

The manner of safeguarding security will depend on the scope of the airport operation, individual destinations, airlines, kind of operation (scheduled, charters or general aviation) and the airport size. In designing the airport protection, it is necessary to consider the following:

➤ protection of the airport perimeter, aircraft on the ground whether in hangars or on movement areas of the airport, operational facilities, stores and terminals

➤ limitation of movement of persons and vehicles into security sensitive areas on the airside

➤ security checks of passengers and employees

➤ control of movement of passengers and separation of arriving and departing passengers.

15.3.2 Airport Perimeter Security and Staff Identification

The standard of safeguarding security on the manoeuvring area and other areas of the airport depends on the reliability of inspection of access to the airport and identification of the employees. It is necessary to emphasise that there is no security system which cannot be beaten. Modern protection of an airport perimeter rests above all in technical installations, which can identify the place and manner of violating the perimeter with high accuracy and reliability. At the same time, violating the obstacle requires a finite time during which it is possible to send a security commando to the place of violation.

On small airports in the third world countries, the outside perimeter of the airport covers several kilometres, which are only very seldom protected by a high quality fencing. On some airports in

Africa, this has even allowed attacks of organised groups on the cargo holds of aircraft while they were standing in the queue and waiting for permission to take off. However, the large open space on the airside of airports allows the threat to be visible and provides in this way enough time to identify the intruder and to respond before they can threaten the terminal area. It is possible only to hope that the security commando has been properly trained and motivated in order to be able to control and defend the apron and the terminal building. Since the number of workers is in most cases low on such an airport, their identification by means of simple cards, possibly with colour differentiation for authorisation of entrance into the specified areas, will be sufficient.

Figure 15-1   DTR 2000 - fencing for areas with high security needs

The protection of the airport perimeter is decisive not only for safeguarding the security but also the safety. Perfect fencing prevents penetration of animals to the movement areas and their possible collision with the aircraft.

A higher security standard is required in large international airports, or at least of some premises in such airports. The required standard cannot be usually achieved by common fencing. In order to secure the more important facilities, barrier type protective systems have been used as well as other types of systems, which function on different physical principles. The quality of security of a system is measured by the probability of intrusion of the protected facility without identification of the place and time of intrusion. The quality of security is one of the most important parameters for judging the level of safeguarding a facility and can be expressed mathematically.

The probability of system intrusion is in particular affected by:

✈ number of false alarms in relation to the unit of length and time (e.g. to 1 km of length in one month)

✈ reliability of the whole system (depends on the reliability of individual parts)

✈ monitoring of the condition of the security system

✈ raising the alarm upon failure of basic functions of the security system.

The intruder can be successful in overcoming the security system, if at the same time four conditions have been fulfilled:

✈ perfect knowledge of the system (mechanical construction, physical principles and software) by the intruder

✈ sufficient time and ideal working conditions

✈ availability of special tools and facilities, which are not normally carried by people

✈ upon unsuccessful first attempt there is a possibility of an unlimited number of other attempts.

When selecting a security system, the system has to be judged also from the point of view of the effect of the environment on its function. Again, there is no perfect system. A system suitable for one locality can be, from the point of view of physical principles on which the system functions, inapplicable in another locality.

✈ **A microwave system** is unreliable on uneven terrain or if there is vegetation (e.g. grass, shrubs) in the detection zone. It can be easily penetrated in the vicinity of transmitter and receiver - without being discovered the intruder will crawl through under the microwave beam.

✈ In the **electric field system** the shape of electric field will change if a large body comes close to it. This change will be recorded by wire sensors. The basic requirements of faultless operation are perfect suspension of the system, ensuring the correct distances between the top wire and the earth and the prescribed stretching of wires. If these conditions are not met, false alarms take place.

✦ **Systems sensitive to pressure** or deformation record the changes in mechanical load on the soil when a person walks above the sensor. False alarms arise by the movements of surrounding trees and poles (upon the effect of the wind) and precipitation phenomena (rain, hailstorm).

✦ **Infrared radiation** (the so called infragates) systems are sensitive to fog, which changes and disperses the infrared beam. Similarly their functioning is affected by dust and by rapid changes of temperature which can condense water on the sensors

✦ **Vibration** or **deformation detectors** can be installed on the finished fence and record vibrations of the fence evoked by intruders. The system is sensitive to false alarms evoked by wind, hailstorm etc.

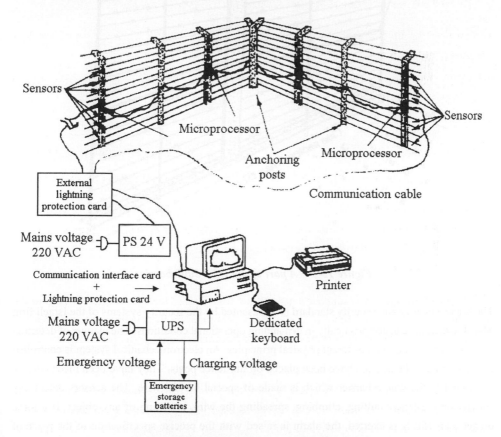

Figure 15-2   DTR 2000 system configuration

It would theoretically be possible to create a perfect, insuperable system by combining various systems. However this is practically not possible. By integration of several systems into one unit the number of components increases and the reliability of the system decreases. The number of false alarms increases and the integrity of the security system, as a unit, decreases. Moreover, by integrating several systems the price of the security provision increases.

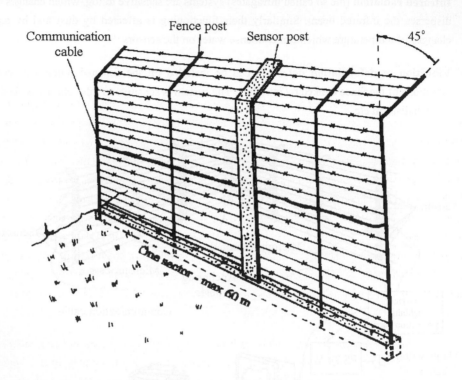

Figure 15-3   A part of the DTR 2000 barrier

The highest quality and security standard is represented by the security systems of the Israeli firm Magal, which were tested practically in demanding operational conditions. The manner of detection of the intruder is based on different physical principles. An electromechanical system is controlled by a computer. The sensors have been placed on the fence posts. They have been mechanically connected to the whole barrier which is made of special barbed wire. The sensors detect any activity on the fence: cutting, climbing, spreading the wire, attaching of any object. If a force bigger than 150 N is exerted, the alarm is raised with the precise specification of the place of intrusion on the operator's monitor. The DTR 2000 system, which is the system with the highest security, achieves a false alarm rate of once in 3 months within 1 km length of fence.

In order to prevent the crossing of the barrier by means of a 'high bridge', it is possible to increase the level of detection by installing an independent camera system monitoring the protected area parallel to and within the guarded line.

Each Magal system contains an automatic meteorological unit, which scans meteorological data. On the basis of measured values the computer sets the threshold sensitivity of sensors, which decreases further the number of false alarms.

15.3.3 Employee Security Procedures

On all bigger airports it is necessary to ensure computer control of authorisation of entrance of workers into specified areas of the airport. The mostly common manner of inspection is the use of magnetic cards. In order to ensure long life of the card despite rough handling, the cards are made of vinyl, in contrast to cash dispensers that are made of polyester. At the entrance point, the worker pulls the card through the slot of the reading device. The authorisation of his or her entry into the given area will be checked and entrance will be permitted or refused. At the same time it is possible to register the entrance or leaving the given area and the movement of the employee on the airport. If security is violated, it is possible to evaluate retroactively the presence of the employee on the airport. The system should monitor also forced intrusion into the guarded areas e.g. through the fire door. It must also enable the inspection of validity of the cards. Cards with limited validity, e.g. issued to part time workers, building workers or concessionaires, must be refused entrance into the protected area after expiration of the validity.

Inspection of the entrance into sensitive areas from the security point of view can be tightened by the combination of the security card with the personal code of the employee, the Personal Identification Number (PIN), which has to be typed on the keyboard after the card has been inserted into the reading device. This will limit the abuse of the card if it is stolen.

Identification of a worker in highly sensitive areas can further be enhanced by the inspection of biometric signs (dactyloscopic signs of the hand, voice identification - voice spectrum upon pronouncing of the password, scanning the iris or face geometrics), comparison of the portrait of the worker scanned by CD camera with the picture stored in the database etc. Some firms offer identification cards which can be read at some distance from the scanner. Such system or cards are substantially more expensive.

Acquisition and operational costs can be reduced by decreasing the number of inspection points into the building. Then it is possible to combine the computer systems for the identification of employees with the control systems of the buildings, which fulfil other functions such as fire alarms, control of the heating, automatic switching off the lights in case of absence of the worker in the workplace etc, i.e. the building management system.

The computer systems for identification of the airport employees and securing the airport are becoming much cheaper and so allow the numbers of security service workers to be reduced. At the same time the costs of maintaining the system increase, while reducing the number of workers can be the source of conflicts with the unions in some countries. In some countries when employing the workers at an airport, both by the airport and by the concessionaires, it may be advisable to screen their previous 20 year history. Should the screening fail, the possible contact of the airport employees with terrorists when, for example, cleaning aircraft, can be made more difficult by random assignment of workers into work groups with the use of a computer program.

15.3.4 Measures in Relation to Passengers

One of the important measures for increasing security is complete separation of arriving and departing passengers. On many airports in the third world countries, the security measures are insufficient and smuggling of the weapon or explosive to the board of the plane is much easier than in the European countries. After arriving from such a country, the terrorist could hand over a weapon to his accomplice, who had already gone through the security inspection. Separation of flows of arriving and departing passengers makes it radically more difficult to hand over a weapon or any other inadmissible material to another passenger.

It is also common to generate profiles of passengers and differentiating the scope of the security inspection. It is obvious that it is necessary to pay more attention to two men coming from a country sponsoring terrorism than to a group of pupils an excursion. Large established airlines have much information in the reservation systems which can also be used for data input for security inspection. If a suspicious passenger or a group of passengers who cannot be assigned to common categories appear on the flight, increased attention can be paid to them. At present there are several databases on the most wanted terrorists in the world. The target should be linking of the databases with the reservation systems so that the attention of the security service workers is automatically drawn to the potential terrorist. It will then be possible to identify the terrorist by means of biometric signs.

Security inspection of the registered baggage was introduced as a countermeasure on the basis on the sabotage on Pan-Am 747 over Lockerbie in Scotland. The aim of the inspection is discovery of the explosive or other banned substance (combustible materials, acids, drugs) in the baggage of the passenger. Performing security inspections of registered baggage means substantial change in the common checking process. The security inspection should be carried out before the passenger is separated from the baggage in order to be able to unlock the baggage and to be present at its manual inspection and also to be able to clarify possible problems. Therefore the security inspection should be executed before or, even better, at the same time as check-in. In order to compensate the deficiencies of the present technical equipment, it was required in some countries, like Great Britain, that 10 % of the overall number of bags be checked manually. In some airports the problem was temporarily solved by establishing temporary inspection stands before the check-in counters. However this solution has all the problems of decentralised security inspections.

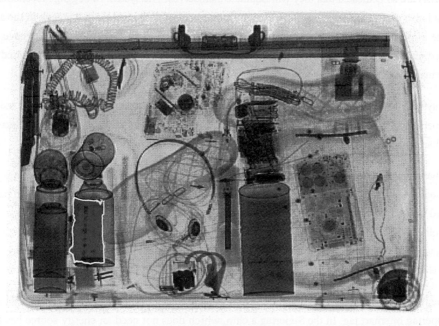

Figure 15-4   X-ray black and white image of a baggage

Another solution is the installation of technically perfect inspection facilities directly in the check-in counter with remote centralised control and signalling of the need of manual inspection of baggage or, alternatively, permission to send off the baggage for loading into the plane.

At Heathrow a system was installed for American Airlines where one worker checks X-rays (E-Scan) from 22 counters. The baggage is inspected by X-ray unit and by gas analyser within the course of check-in process. The total time of the check-in process is not extended and also the requirements on the number of security workers are lower.

The ultimate requirement is to require 100 % inspection of all registered baggage either by technical devices or in combination with manual inspection. Meeting such requirement means extending the check-in process, and higher costs and space requirements. This would substantially disturb the airport functions and in this way the terrorists would manage to achieve one of their main targets, namely to disrupt substantially the running and economics of the country.

Another requirement on the baggage inspection is to ensure baggage reconciliation of the registered baggage with the passengers. It has to be ensured that all passengers that have handed in their luggage during the check-in process have also boarded the appropriate aircraft. Only a few terrorists are so dedicated that they will liquidate themselves. The system of baggage reconciliation should also ensure a permanent survey on the movement of passengers and baggage in the areas of the airport and the information as to which part of the aircraft hold space the baggage was loaded, in case it is necessary to search out and unload the baggage. It also has to be ensured that after inspection there is not a possibility to replace it or to put anything into the baggage.

The system of baggage reconciliation has two principal shortcomings. In fact there are suicide bombers, who will bring the bomb on board themselves and detonate it, and also there are persons that had a bomb planted into their baggage without their knowing it.

The baggage reconciliation is provided by computer comparison of databases of baggage tags with the boarding card of the passenger, which were made upon check-in. The baggage tag can have a record with a bar code, which can be read by laser sensors. Some systems use a baggage tag record as well as a boarding card record. The code is at the same time used as information for sorting of baggage. Before boarding the plane or upon passing the chosen places the boarding card will be inserted into the reading device and in this way the movement of the passenger will be monitored.

In the future the bar code will probably be replaced by an electronic baggage tag, e.g. of the type – Supertag or smart tag. In the Supertag a chip, which does not need an energy source has been installed, with a flat antenna. The chip was originally developed for supermarkets. The scanner can read the information within the distance of several meters at a speed of 50 objects per second.

Baggage inspection can be performed manually or with the use of technical facilities, at present particularly with the use of X-ray units.

Manual inspection of baggage has several advantages over the X-ray inspection. E.g. identification of the majority of objects in the baggage is easier compared to the interpreted display on the screen, it has lower initial costs, and thoroughness of inspection depends primarily on the spent time and so it is possible to react promptly to the external conditions.

On the other hand inspection with the use of X-ray units has several advantages over the manual inspection. It is easier for the X-ray unit to discover secret partitions in the baggage or objects hidden inside other things, it enables higher productivity and hence decreases labour requirements. The passenger's privacy is less disturbed when his or her personal belongings are being inspected. At the same time it is difficult for the potential terrorist to predict how successful he can be in hiding the weapon against the X-ray inspection.

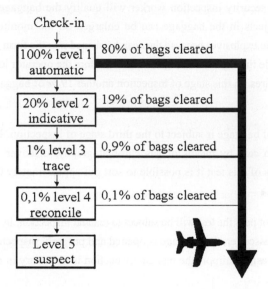

Figure 15-5   Automated 5 stage - 100 % hold baggage screening

Automation of security inspection with the use of technical devices has a number of advantages in comparison with the inspections that are executed personally, by security service workers. The machine is different from humans in that it does not decrease in efficiency and reliability when repeating monotonous activities. The machine also does not make mistakes. On the other hand the technical facilities are only the tools that speed up the security inspection and make it easier. In order to ensure effective inspection the operating personnel must be properly trained and acquainted with the facility in order to be able to use it effectively.

In the European Civil Aviation Conference (ECAC), the 37 member states have set a target of 100 % screening of international hold baggage by 2002.

All international airports are going to have to install 100 % hold baggage screening. If automated, the baggage inspection may be designed as a 5 stage process. At the beginning of the process there is a computer controlled X-ray unit that automatically sorts out the baggage. The suspicious baggage is searched for according to the weight of articles in the baggage, affinity of shapes of the objects with suspicious objects, and the atomic weight of substances. During this test, the 80 % of baggage which is not suspicious, is sorted out and can be immediately loaded into the aircraft.

Approximately 20 % of baggage which does not pass the automatic inspection is subject to further inspection by X-ray unit supplied by the firm Vivid. The X-ray unit marks the possible explosives with a red colour. The security inspection worker will qualify the baggage on the X-ray unit monitor. Suspicious objects in the baggage can be enlarged on the monitor. Sometimes it is difficult to identify plastic explosives because they can be shaped into the form of common objects. It is more difficult to hide timing devices or primers. Therefore the operator looks particularly for devices connected by wires. In this stage of inspection another 19 % of baggage is sorted out and sent off to the aircraft.

Each suspicious piece of baggage is subject to the third stage of inspection. In this stage the gas analyser is used, which can, by means of gas chromatography, discover the traces of some explosives. On the basis of this test it is possible to sort out approximately 0.9 % of suspicious baggage.

The baggage that does not pass the test will be subject to manual inspection. In a separate room and in the presence of the passenger, the baggage is opened and properly inspected by an explosives expert. According to Glasgow airport the manual inspection is necessary in only one of 5000 to 7000 pieces of baggage.

A different approach has been chosen by the Israeli airline El-Al, which in addition to technical facilities relies on ascertaining the passenger's profile in a conversation with the security inspection worker, together with proper manual inspections. In the course of the inspection the passenger is asked questions separately by two workers, the answers being compared and evaluated. Answers are of the same importance as the reactions of the passengers. A precondition of their correct interpretation is high professionalism and experience of the security service workers. However, the conditions of El-Al company are unique. Particularly close co-operation of security departments with state administration and their direct link to the information service is necessary. El-Al has a relatively few international flights, so it is possible to provide inspections by elite workers at each point. The precondition of this approach is not only extensive support of the state but also top level professional training and high motivation of the workers.

15.4   DETECTION OF DANGEROUS OBJECTS

As has already been mentioned, inspection with the use of technical facilities has a number of advantages, which make the security inspections easier and speed them up and at the same time limit the possibilities of human factor errors. When selecting the technical facilities it is particularly necessary to take into account the airport size, number of passengers in the peak hour, passenger and flight profiles, level of wages, acquisition and detection costs.

Detection devices for discovering explosives and other inadmissible objects have to meet many criteria. They must be able to discover weapons and explosives, which are used in the military or in the civilian sector. The sensitivity of the device must not be affected by the location of the explosive or the type of container and, moreover, the device has to function with a minimum of subjective human input. The price of the device and its peak capacity are also important. Several devices have been developed, which meet most of these criteria. These include metal detectors, X-ray units, neutron activation analysis, gas analysers, vacuum chambers and use of trained dogs.

Metal Detectors

Modern metal detectors can discover weapons made also of non-magnetic materials, composites or metals with low electrical conductivity.

X-ray units

X-ray devices were first used at airports in the USA at the beginning of the seventies. They have become an effective device for discovering potential hijackings, who at this time represented the greatest danger. The first X-ray units were not sufficiently sensitive and the display of the baggage on the monitor in real time did not provide enough time for the operators to recognise dangerous articles in the baggage. The picture was not sufficiently clear. In addition to other things the X-ray capabilities were characterised by two, to a certain degree different parameters:

✦ resolution, indicated by the thinnest steel wire that an X-ray unit is able to display on the light background

✦ penetration rate, measured as the thickest steel plate that an X-ray unit is able to radiate through without having the plate displayed as a completely black object.

However these criteria do not represent a standard accepted all over the world.

X-ray units have also to meet health standards concerning radiation leakage. According to the standards valid in the USA, radiation leakage must not exceed 0.5 milli Roengen per hour(mR.h^{-1}). In modern devices this leakage usually does not exceed 0.1 mR.h^{-1}. X-ray system must be safe for photographic film, magnetic carriers of data or data in electronic notebooks.

In 1980 the firm EG&G Astrophysics introduced Linescan System I, the first digital system, which used as sensors a series of silicon diodes, producing black and white images. The system gave a high quality of display and high reliability. Sensitive sensors enabled the radiated output to be decreased so that the device was safe for a photographic film. The device has been further improved, individual objects in the baggage being displayed in pseudo-colours depending on the material's thickness and density. The principle of a modern X-ray unit is shown in Figure 15-6.

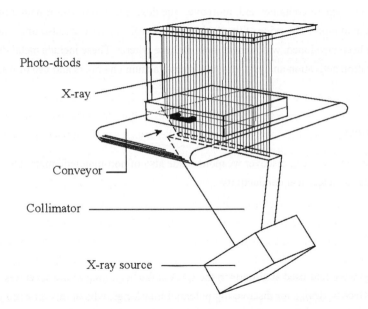

Photo-diods

X-ray

Conveyor

Collimator

X-ray source

Figure 15-6   Principle of modern X-ray unit

In 1988 EG&G Astrophysics introduced the first X-ray unit functioning on the principle of radiation of two different energy outputs (Dual Energy Systems). This type of X-ray unit is known by the abbreviation E-Scan. An X-ray unit with the same technology was developed by the firm Heimann and is being offered under the name HIMAT. E-scan is capable of distinguishing organic and inorganic substances by their atomic numbers. Objects with atomic number below 10 are classified as organic and on the monitor they are displayed in orange / brown shades. Explosives and narcotics have these low atomic numbers. Plastics and explosives such as Semtex or C-4, which also have a low atomic number, are displayed by orange warning colour on the monitor, which attracts the attention of the operators. Objects with atomic number higher than 10 include metal objects (e.g. weapons). These are displayed in shades of blue. Very thick objects that cannot be penetrated by low energy radiation, are displayed in green. In this way the attention of the well trained operator is drawn to the suspicious objects, the baggage then being manually searched.

Another type of X-ray unit, which is offered by the firm Heimann under the name HI-CAT (Computer-Aided Tracing) uses a computer to search for suspicious objects. Shapes of dangerous objects (guns, knives etc.) have been saved in the computer memory, these being compared with the objects in the baggage. The attention of the operator will be drawn to the object by the computer.

An X-ray unit which gives a three dimensional display of baggage is being offered under the name Z-Scan. The three dimensional display is achieved by associating two emitters in one X-ray unit. The first radiates the baggage from the side and, with a time delay, the second cross-ways from the bottom. The results of both X-ray units are processed by a computer, which assigns a division of weights in vertical sections through the baggage. From individual sections a three dimensional picture of the baggage is generated. By means of an algorithm the location of atomic weight of plastic explosives is identified and the attention of the operator of the device is automatically be drawn to them. The inspection of one piece of baggage does takes no longer than 5 seconds.

An X-ray unit from the firm Vivid Technologies - USA, uses, for identification of dangerous objects in the baggage, the technology of the Hologic firm's computer tomography, which was originally used in health care. As in the case of E-Scan, the baggage will be gradually radiated by 70 kV energy and then by 140 kV. The results of both tests are processed. By comparison with the database, the system is automatically able to determine suspicious objects. They will be displayed in red and the attention of the operator will be drawn to them. The advantage of this device is that it can also work fully automatically and suspicious baggage will be sorted out so that the operators can concentrate on the suspicious baggage.

The majority of modern X-ray units have functions that substantially simplify the work of operators and remove the shortcomings of the first X-ray units with imperfect television monitor displays. Digitalised videorecording allows the object to be searched for any period of time with further processing of the picture, e.g. enlarging of one part of the baggage. Emphasising the edges of individual objects and wires (Edge Enhancement) makes it easier to discover electronic primers. Identification of objects behind very thick objects (lead plates) is made easier by gamma emphasising (Gamma Enhancement), which makes the picture lighter or darker.

A recent improvement, which was introduced to the market by the firm EG&G Astrophysics in 1994, is replacement of traditional photodiodes by a combination of optical fibres and a CCD camera. In comparison with the photodiodes this technology has low interference and substantial improvement of the picture quality even with a high speed conveyor belt. While with traditional X-ray units the speed of the conveyor belt can be around 0.25 m.s^{-1}, with the new Super E-Scan-200 a high-quality picture is provided with conveyor belt speeds above 1 m.s^{-1}. This substantially increases the capacity of the system of baggage inspection.

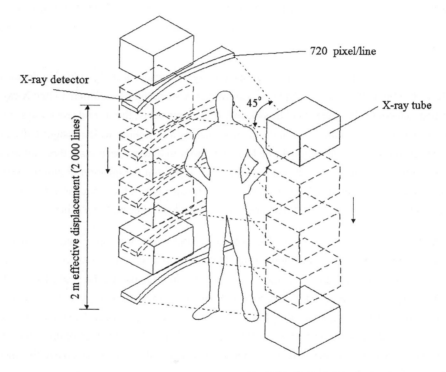

Figure 15-7   Principle of FOSCAN Fibre Optic Scanner

Modern X-ray units equipped with a microprocessor are more reliable and require less maintenance. They allow automatic calibration and inspection of the device with identification of faults.

The advantage of X-ray units in comparison to TNA (see below) is their price. A modern X-ray unit, which is capable of distinguishing organic and inorganic substances costs around 55 000 US dollars in comparison with approximately 1 million dollars for TNA.

For the inspection of persons before entering areas with special security needs, it is possible to use a soft X-ray unit made by Foscan. The high sensitivity of the device and the system of scanning, which synchronises the movement of the emitter and the scanning device substantially decreases the level of radiation. The total dose of one inspection represents approximately 3 % of the radiation of a classical X-ray unit and corresponds approximately to the dose to which the human would be exposed naturally during the period of 24 hours at an elevation of 3 000 m.

Neutron Activation Analysis (TNA)

The method of Neutron Activation Analysis, known as Thermal Neutron Analysis (TNA) has been developed in the USA. By means of TNA it is possible to find out the composition of any object in the baggage. The physical principle has been known for a longer time, however its practical application has been used only recently.

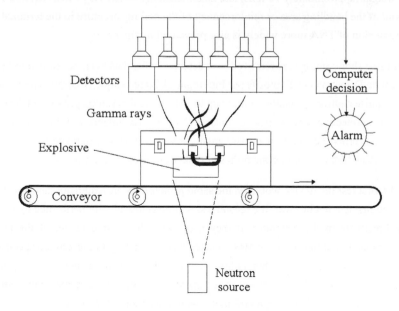

Figure 15-8   Principle of TNA

There is a source of low-energy thermic neutrons located in the device. When passing through the substance there is a high probability that the thermic neutrons are caught by atoms of the substance. Then the atoms of the substance change into isotopes with an energy surplus for a very short time (fractions of a second). On return to the original unexcited or low energy form, the surplus energy is released and radiated in the form of high energy gamma radiation, unique to each specific substance. By evaluating relative intensity and energy of radiation it is therefore possible to specify the presence, location and amount of specific elements in the inspected object.

Explosives are characterised by high contents of nitrogen and a certain relation of other light elements - oxygen, carbon and hydrogen. If individual elements in the objects of the baggage are compared to the known composition of the elements of the explosive, the presence and type of the explosive can be stated with high accuracy. In a similar way the appearance of other substances, e.g. narcotics can also be ascertained.

The first device which works on the TNA principle to achieve a reliability of 95 % was developed and produced by the American firm Science Application International. Six TNA devices were located on airports in the USA and in Great Britain. However the tests of the device were not satisfactory. The problem was the small capacity of the device (approximately 10 bags a minute), its high weight (approximately 10 tons) and its dimensions, which very often required changes in the lay-out of the handling area or reinforcement of the ceiling structure in the terminal building. The introduction of TNA more widely is also prevented by the price.

Some of these shortcomings are removed by a detector which has been developed in France. It is Equipment de Détection des Explosifs par Interrogation Neutronique (EDEN). The nucleus of the device is a pulse neutron generator emitting up to 100 milliard neutrons per second. The generator is located in a protective polyethylene or concrete chamber. The intensity of the current of neutrons ensures the specified accuracy of measurement, however does not damage the contents of the baggage such as films, computer memories or diskettes, electronics, articles of food, etc.

During the first test the device works on the same principle as an American detector. Should the result of the first test not be clear, the second test will be automatically performed by fast neutrons emitted directly from the generator. If they meet with the atomic nuclei of the investigated substance a reaction of non-elastic diffusion takes place, radiating in the gamma spectrum, which is different from the first test. This additional information makes it possible to eliminate the uncertainty of specifying the contents of the baggage. The baggage is automatically sorted out and either it is cleared or, if still regarded as a suspicious item, it will be subject to additional inspection. The main advantage of the French system is prolonged life of the neutron tube and the capacity of 1 400 pieces of baggage per hour.

The capacity of the device, in addition to the price, is a decisive criterion in selection of technical facilities on large airports, given that a Boeing 747 has on average 700 to 800 pieces of registered baggage on the board.

Gas Analysers

Terrorists prefer plastic explosives. Some of them, such as Semtex and C-4, are more effective by one third than TNT and this is twice as effective as dynamite. The basic components of all plastic explosives are either RDX (cyclotrimethylentrinithramin) or PETN(pentaerytrytol tetranitrate). Highly explosive substances pass almost immediately into gaseous state after explosion, the speed of detonation of C-4 being 8 052 m per second.

The plastic explosives were developed during the second world war, because they are more effective but also resistant to the environment as they can be detonated only by a primer, and also because they can be formed. By precise dosing of additives and shaping, it is possible to give them almost any appearance, from bricks, leather shoes to childrens' toys. Another advantage of plastic explosives, for example Semtex, is that they have practically no smell and their emissions can be identified only with difficulty.

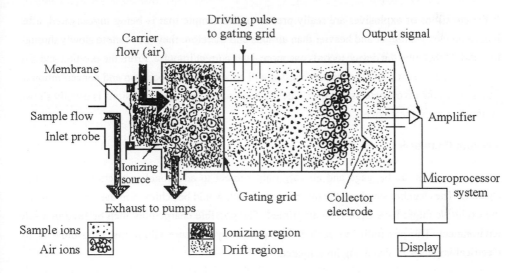

Figure 15-9   *Principle of gas analyzer*

Therefore the member states ICAO adopted in March 1991 the convention on marking plastic explosives to aid their detection. In the course of production the plastic explosives have to be marked by one of the four agreed substances. These are ethylene glycol dinitrate; 2,3-dimethyl-2,3-dinitro-butane; para-mono-nitrotoluene or ortho-mono-nitrotoluene. The aim is to increase the level of emissions of material so that they can be more easily identified by common gas analysers and by dogs. Unmarked explosives are limited to use for military and police purposes.

The devices for identification of explosives are used as a supplement to other technical facilities, e.g. to X-ray units or as independent devices. In order to identify explosives on the passengers, the gas analysers can be built into, for example, a revolving door. The device can further be equipped with a magnetic card sensor for inspection of the entry of employees into especially high risk areas (e.g. air traffic control centre). Manual detectors can be used for inspection of baggage.

Some of the detection devices work on the principle of gas chromatography (Figure 15-9). Emissions from the investigated object are sucked by a pump into the first chamber of the detector, where they will hit the membrane of the silicone rubber. The emissions of the explosives pass through the membrane into another chamber, where they and also the air molecules are exposed to a weak source of radioactive radiation, which ionises them and the molecules acquire an electrical charge. The ionised molecules are then driven through a tube with an electrical field pulsating with the frequency of 50 Hz, after which they hit a scanning electrode, which records them in the form of a weak electrical current.

If the emissions of explosives are really present in the sample that is being investigated, after ionisation they get bigger and heavier than air ions and therefore they move more slowly through the tube. At that time the device records one more, later current impulse. From the position and size of the second current peak, the microprocessor will specify the composition and concentration of emissions of the investigated substance. The relative concentration of emissions is usually shown on the display of the device.

Vacuum Chambers

At some airports the baggage and air cargo are subject to additional inspection in a vacuum chamber. The chamber simulates the low pressure which would be achieved in the aircraft in which the explosive might have been vacuum primed. The possible charge may then be fired in a safe environment. The possibility to explode any primer by pressure vibrations, ultrasound or by electrical impulses is also being investigated.

Dogs

Using trained dogs is still a very common way of discovering explosives. Dogs are very fast, work effectively and are able to sniff out what most of the devices cannot discover, such as clear plastic explosives and trinitrotoluene. Another advantage of dogs is that they can penetrate also places where the human would never get.

The disadvantage of dogs is that they cannot work by themselves and independently. Also, depending on the conditions applying in a particular country, there can be high costs for taking care of them. They must be managed and trained by an experienced dog handler. A dog may lose concentration, lose interest or its attention can be distracted by another object. In spite of that, it carries on performing the trained activity. Another problem is that the smell of the searched object can be covered, either intentionally or not, by other strong smells e.g. by solvents, ammonia or fuel.

Therefore it is best to use the dogs more as an additional procedure, e.g. for inspection of aircraft, baggage and objects when there is a danger of bomb attack. The dogs are less suitable for regular inspection of baggage.

15.5 SUMMARY

Each airport must have an airport emergency plan, the extent of which corresponds to the airport size and its importance. In addition to other things it must specify not only the duties of security departments but also of individual employees for different kinds of emergency situations. Examples of such situations can be fire, attacks on the airport and its premises (stores, fuel farms, air traffic control, apron etc.), planes or hijacking of planes or landing of a plane with the terrorists on board. The plan has also to identify the emergency staff that will proceed according to the planned procedures.

A higher quality of technical facilities decreases the probability that the saboteurs can get the explosives or weapons on board aircraft by traditional means. However this does not mean that they will give up their activities. On the contrary it is possible to assume that they will look for other ways to achieve their objectives. Therefore it must be ensured that all routes to the aircraft are protected. It is no use installing facilities for detection of explosives for millions of dollars if the security measures are such that it is easy to get to the apron and to the planes without being detected. Equally it is a waste of money if the operating personnel are not sufficiently trained and motivated. In other words high quality technical facilities are welcome, however it must not be forgotten that they can be effective only when fully integrated into a sophisticated security system.

16

LANDSIDE ACCESS

Tony Kazda and Bob Caves

16.1 ACCESS AND THE AIRPORT SYSTEM

The primary advantage of air transport is speed, particularly for long haul where it has completely supplanted shipping, but also for those short haul trips where it is in competition with surface transport. However, the average speed is reduced by the ground portion of the trip. The trip does not start or finish in the airport, but at home, at the hotel, at the workplace etc. The passenger is just as concerned to reduce time on the ground as in the air part of the trip, and just as annoyed by any delay, whether in the air, the terminal or on the way to and from the airport. The total time of transportation 'from door to door' is decisive for the passenger. The attractiveness of an airport markedly decreases if the time of access by surface transport exceeds a certain maximum time. For short haul trips this might be as short as 30 minutes, while two hours or more might be acceptable for intercontinental trips.

Transportation time to the airport is just one of the three most important factors effecting the decision process of the passenger for a particular airport, the other two factors being price of the flight ticket and number of flights (frequencies) offered. The quality of surface transport affects the size of the catchment area of the airport in the competitive market between several airports. Surface transport quality and quantity should not need to become a limiting factor of the development of air transport on the airport if plans have been properly developed. Yet, paradoxically, it may be the very success of increased access traffic that limits an airport's growth, if the environmental impacts of the ground transport are too severe. The problem remains that in some cases there is no co-ordinated approach to the planning for an airport and that for the town and the region that it serves, particularly with respect to transport. Yet the peaks of traffic to an airport often mirror the peaks in the local traffic. Sometimes the responsibility for planning and operation of the various modes of transport have been divided among several departments.

Table 16-1   Distance of some airports from city centres and connection times by public transport

City	Airport	Distance [km]	Connection times (public transport) [min]
London	Gatwick	43	31
Chicago	O'Hare	35	29
Paris	Roissy	28	35
New York	J.F. Kennedy	27	38
New York	Newark	25	32
London	Heathrow	24	55 (15)
Tokyo	Haneda	19	15
New York	La Guardia	13	18
Frankfurt	Rhein Main	10	12
Paris	Orly	10	31

NB ( ) by Heathrow Express

From the beginnings of air transport the airline companies laid great emphasis on the provision of transportation of the passengers to and from the airport. Many passengers had no other means of transportation. The airports have been, contrary to the railway stations, located outside the towns they served. Therefore the airlines found it quite natural to provide transport for airport access, mostly by bus. The first connections of the airport and the town by rail were built in Berlin and London at the end of the thirties. After the Second World War, as a result of mass development of passenger car travel, there was an increase in the use of the car for transportation to/from the airport, and the need for improvements to the road system to allow convenient access.

16.2 SELECTION OF THE ACCESS MODES

A gradual increase in the share of the available high occupancy modes of transportation normally occurs as the airport grows, and this change should be encouraged on the grounds of environmental impact and balanced capacity. Car trips will always predominate at small airports. Growth in the share of high occupancy vehicles as the traffic increases will normally take the form of public (mass) road transportation at moderately sized airports, while high capacity rail transportation should have a substantial share in the large airports. Mass transportation should have a role by the time an airport reaches 2 million passengers per year. Many factors make the car the preferred mode of access, including the low marginal cost, the convenience for carrying bags and family groups, and the instant availability. It is therefore not easy to get people to move to high occupancy modes even if they are competitively priced, frequent and form part of a transport network that allows access to the complete catchment area. The airport administration has a new role to ensure that the passenger can get to and from the airport quickly, easily and simply so that they do not miss their flight. If the management does not do this, some potential passengers will be lost to other airports or other modes of transport. Surface transport must be considered as a part of the 'product' of the airport. Managers have to work to fulfil the airport's access needs by encouraging the local authorities and transport operators to respond by investing in roads and operating services.

Theoretically it would be possible to ensure the change in the share of individual modes by making the mass transportation more attractive or by making the private transportation less attractive by imposing road tolls, high parking fees etc. However, individual groups of people accessing the airport, who will mostly be passengers and those accompanying them, employees or visitors to the airport, will all rank the factors differently. The following factors can be identified as affecting the selection of the mode of transport:

✈ the availability of the mode

✈ the distance of the airport from the town

✈ length of the individual elements of the transportation process (waiting, time to access the mode of transport, transportation time, time from the mode to the airport check-in)

✈ standard of comfort and quality of transport, which includes ease of use, number and quality of seats, handling of baggage, number and difficulty of transfers en route, possibility of secure parking

✈ reliability of transport

✈ total generalised cost of transportation (parking fees, value of time etc. must be included as well as fares or marginal cost of using private transport)

✈ other factors such as personal safety, privacy, flexibility.

Research shows that the decisive factors for passengers are: price, transportation time, number of changes and baggage handling. For workers, particular concerns are flexibility, availability and personal safety for shift patterns out of normal hours.

16.3 CATEGORIES OF SURFACE TRANSPORT USERS

Airport access is needed not only by the airline passengers but also by other important groups of users. The number of employees who daily commute to the airport represents one quarter to one half of the daily number of airline passengers medium size and large airports. In addition there are the accompanying persons (meeters and greeters) and the visitors to the airport, this category amounting to 5 or 10 per cent of the total. The roads and public transport networks also have to cater for the needs of the local non-airport traffic.

It is necessary to emphasise that there are also other factors which distinguish the airline passengers from other groups of surface transport users that have often a decisive effect on the selection of the kind of transport provided, and these can be different in each case. It is, for instance, necessary for business people to have fast reliable transport, leading to the development of the Heathrow Express. Therefore it is necessary to distinguish categories of passengers using surface transport, characterised by the factors such as:

✈ the reason of the trip to the airport

✈ type of flight (scheduled – charter, short haul - long distance flight)

✈ duration of the stay

✈ social and economic factors (income, age, occupation, size of household, car ownership)

Each category will have different requirements.

16.4 ACCESS AND TERMINAL OPERATIONS

The characteristics of the various categories that determine access requirements also affect the requirements for the terminal. Those who want a fast and reliable access trip also want to move quickly through the terminal with as few surprises and delays as possible. They will perhaps only expect to check in at the departure gate 10 minutes before the flight, having been assigned their seat when making the reservation. In contrast, those with no time pressure and with little experience of air travel will accept a relatively slow coach journey to the airport and a long wait in the terminal in exchange for the assurance of not missing their flight. Indeed, the tour operators encourage these passengers to arrive early, and schedule their coaches similarly. It will be particularly important to them if they are travelling on inclusive tour or on discount fares not to miss their flight, since their ticket will not be interchangeable. They may be in the terminal for three hours or more, and will require many more facilities for food and relaxation.

A similar disparity exists between long and short haul passengers. The airline will need the long haul passengers to be checked in early because there are more formalities to be completed and there is more luggage to be checked in. Also, since fewer airports offer long haul flights, the average access time will be greater and the passengers will need more time to organise themselves before flying. The proportion of leisure travel on long haul is high, so a high percentage of the passengers will have discounted non-transferable tickets and will have the same desire not to miss their flight as the inclusive tour passengers. So, depending on the type of passenger, each will have their own punctuality targets for the access trip and will adjust their start time to meet it. If they are inexperienced travellers, they may well experiment with the access trip in order to get some idea of how long it will take and how to navigate along it.

The actual dwell time in the terminal depends on the actual time of the complete access trip, including finding a parking space and transferring to the terminal, and the planned margin of safety together with the reliability of making the trip in a given time. The actual time for a trip will depend on the distance, the mode used and the traffic conditions. For each mode, it is likely to be normally distributed with a variance that is strongly related to the average time. If passengers have a good sense of the variance, they will leave a margin of safety that will get them to check-in at the prescribed time with a given level of confidence, say 99 per cent. Those with longer journey times will then, on average, spend longer in the terminal even though holding the same ticket type for the same flight as passengers with shorter access times.

Similarly, access trips on modes or routes with low reliability will have a high variance. This will tend to be known to passengers, and they will again adjust their behaviour to achieve the target punctuality. Thus passengers on low reliability access modes or routes will again spend longer in the terminal, since they will, on average, have arrived earlier. If passengers do behave in this way, and if they value their time, they will switch to a more reliable mode according to weather conditions and time of day.

In any case, the terminal should designed in terms of space and facilities to meet their needs. The correct provision of space will depend on how and where the passengers wish to spend the slack time. Shopping has now become an accepted use of slack time, and in some cases is a deliberately planned activity. However, the processing may also be affected by the knock-on effects of access considerations, particularly if the design uses simulation of representative peak aircraft schedules and distributes the passenger behaviour around them.

16.5 ACCESS MODES

16.5.1 Passenger Car

The high standard of living in developed countries means that cars are widely used for transport, and particularly by air passengers who tend to be richer than average. The car may be owned or rented, the latter being used by passengers whose trip started at the other end of the route and whose local trip origin is a workplace or hotel. Passengers prefer cars because of their flexibility and comfort. The additional factor for the business person is the short door-to-door time. The less well off also prefer cars because they are perceived to have the lowest marginal cost. The real marginal cost of transportation by one's own car can be relatively low, but the parking fees can be high. In some cases tour operators or hotels will reimburse the parking fee to the passenger, or they may provide coach transport.

The high share of private car transport causes growing problems to airport administrations. It is necessary to build new parking places and access roads, and to increase the size of the drop-off zones in front of the terminal building. Constructing multi-level car parks in the vicinity of terminal buildings is costly and takes up space that could be better used for commercial purposes. If the parking places are provided in the remote areas of the airport, in most cases it is necessary to provide courtesy transportation to the terminal building for the passengers, which increases the operational costs. On the other hand, revenues from parking places represent an important part of the revenues of most airports, particularly in America. At the biggest airports the revenues from parking can be similar to the revenues from landing fees. Most of the airports solve the dilemma of the distance of the car-park from the terminal building by dividing the car parks between short term high price spaces in the vicinity of the terminal and the long term remote car parks with a lower charge per day. The share of the number of places in the long term and short term car parks depends on the space available at the airport. A rough estimate of 1000 spaces per million passengers per year may be used, but the total and the split between long and short term parking will vary depending on the characteristics of the traffic in terms of short to long haul flights and the business to leisure passenger. Space in the central part of Heathrow airport in the vicinity of the Terminals 1,2 and 3 is very limited and the rate per day in the short term car parks is approximately four times higher than in the long term car parks. The older airports with little space for further expansion have the biggest problems with building the access roads.

The second important parameter of airports with a high proportion of passenger cars for access is the length of the kerb in front of the departure concourse. Particularly at airports where the majority of passengers is driven there by another person, called 'kiss and ride', the length of the kerb in front of the departure concourse can be a critical point at peak times as the same central terminal area continues to handle the extra traffic by serving more aircraft gates. The rule of thumb for initial design is 100 m of kerb per million passengers per year. The common solution is a double level design of the departure concourse, or operational measures such as speeding up the unloading of passengers in front of the terminal with the aid of traffic wardens. It is also possible to organise the traffic in front of the terminal building in two or more lanes. However this increases the possibility of accidents or incidents when passengers are crossing the lanes to the terminal.

Figure 16-1   Kerb in front of the departure concourse for passenger drop-of
(photo A. Kazda)

A large percentage of private vehicle access trips is not only likely to lead to congestion; it also has a negative impact on the environment. At most airports the ground trips associated with the airport generate a greater share of air pollution than the aircraft movements. At small airports access almost always depends on private transport. These airports' flights consist in the main of short and medium distance routes by regional planes providing spoke connections to hub airports or direct regional point to point transportation. The surface transport must support this system by providing highly reliable and quick transport 'from door to door'. In these lower density situations this can only realistically be provided by the private car, so it is not surprising that it represents the major means of access for all segments of users. Mass transportation to small airports is usually not economically feasible.

16.5.2 Taxi

The taxi has similar requirements for the design of the kerbside in front of the terminal as the private car. However, its kerb occupation time and parking requirements are considerably smaller, the latter consisting of a pool for immediate service and a longer term zone where taxis can wait their turn to be called forward. It would be possible to consider the taxi as an almost ideal means of transport for an airline passenger. It is reliable, comfortable, ensures the direct transportation between the trip origin directly to the kerb in front of the terminal building and substantially reduces luggage handling problems. The high cost relative to the private car is partly mitigated by avoiding the need to park, and, relative to public transport by the ability to share the cost among a group of passengers.

Although the airport administration is not directly responsible for the operation of taxis, the bad impression formed from low-quality or poor value services of taxis have an impact on the overall image of the airport. Therefore the airport administration should lay down criteria for the acceptable operation of taxis. Important factors are:

✈ to ensure the number of taxis meets the demand, particularly at night and in the time when the mass transportation is not available

✈ to ensure the high quality and fair price of services

✈ to ensure security issues in some countries and to discourage unofficial operators.

Shortage of taxis can occur particularly at smaller airports when two flights are arriving close together.

The problems of shortage of taxis and quality of service are mutually interdependent. This is often solved by awarding licences to serve the airport for a limited period of time. Many airports, like Praha – Ruzyně, issue licences against a fee. Other airport administrations, like Schiphol, issue licences free of charge on the basis of regular evaluation of the quality of service.

A disadvantage of taxi, and also of private car, is access to the airport in peak times, where the journey may last longer than by coach which may use special priority lanes or by rail transport.

16.5.3 Minibus

In some countries the transportation between the airport and the hotels or other main traffic generators in the town is provided by minibuses. The minibus, or van, may run according to a time table or wait until it has been filled. It is usually cheaper than a taxi. Some hotels provide courtesy minibuses free of charge to and from the airport. Minibuses represent a blend of public and private transport. They offer some of the door-to-door advantages of a taxi and, at the same time, more security, comfort and speed than a scheduled bus service. They have a high daily utilisation, though they often depart with few seats occupied because, in order to be attractive, they have to be frequent. This can result in a large environmental impact per passenger and congestion at the kerb. This is despite the fact that they take only slightly more space at the kerb than a taxi and substantially less than a bus. An additional problem with those minibuses operated by private companies, as opposed to sponsors like hotels, is that competition encourages them to occupy the kerb as long as possible. This needs strict policing.

16.5.4 Bus

Several types of bus service to airports can be distinguished.

Normal scheduled services of the local metropolitan authority are used more by employees than passengers because of their frequent stops and poor provision for luggage. However, in countries where the use of public transport is high, as in the countries of the former eastern block, municipal buses may be more used for airport access by passengers who are locally resident.

Local shuttle coaches dedicated to airport passengers, for example the Washington Flyer, are more costly, more comfortable and faster than the scheduled buses. However, they usually link the airport only with the municipal rail or bus terminals and the major hotels. They work well for visitors, but are less convenient for locally based travellers who are more likely to want to start or finish their trip from their homes in the suburbs.

Longer distance scheduled coach services compete with rail, taxi, private car or even air access to hub airports from more distant cities whose own airport does not offer the same range of air services. They overcome the need for the change of mode that is required on most rail systems to complete an airport access trip, but tend to be slow, infrequent and, as with the shuttle coaches, they may suffer from delays due to road congestion. They are mostly run commercially by a private company, but sometimes an airline operates the service primarily for its own passengers and may even check bags through to the trip destination.

Coaches for passengers on inclusive tour holidays, particularly those using charter flights, are organised by the tour operators. Some operators provide a pick-up service at the beginning of the holiday, but the main use is to convey passengers from the holiday airport to their accommodation. They are usually organised into a hub and spoke operation, where a series of flights arrives close together, and the passengers are assigned to whichever one of a bank of waiting coaches is to call at the required hotel. The coach parks for this type of operation at the major holiday airports like that at Palma in Majorca have spaces for hundreds of coaches. The theoretical efficiency of this process is often not achieved because of flight delays.

Another example of interconnection of air and bus transport is the system created by National Express Group plc. They own East Midlands airport in Great Britain and also the National Express interurban coach company. The airport has no direct connection to rail transport, so a network of coach services has been created with good and reliable connections, which is an important marketing tool in the competition with the nearby airport in Birmingham.

Both the National Express and the tour operators' models for interlinking air and coach services are examples of private initiatives in line with policies in the US and Europe to provide 'joined-up', or 'integrated' public transport and thus create a more sustainable transport system.

16.5.5 Railway Transport

The recent political push for a sustainable transport system, combined with road congestion and frequent delays, has caused a renaissance of public transport in accessing airports. Many airports are supporting increased use of rail transportation, both to serve the nearest cities and also to increase their catchment, to the extent that airport expansion has been made conditional on achieving targets of up to 50 per cent of passengers using public transport.

It has to be emphasised that rail transportation can be effective only for connection between large towns and large airports and has to be complemented by other kinds of transportation. In designing the airport and planning its development, the transportation to and from the airport has to be considered as an integrated system, including transportation by passenger cars, taxes, rental cars, buses, coaches and railway transport, not forgetting also the use of marine transport and helicopters where appropriate. Theoretically, rail transport allows a good connection to all parts of the catchment area and can substitute for other kinds of transport. In fact the decisive factor is not the extent of the network but the feasibility and reliability of the connection. The disadvantage of rail access is that it usually connects the airport only with the town centre, while most of the passengers do not start their trip in the town centre. First they have to travel from their permanent residence to the town centre and only then to the airport. When travelling by car the passenger can avoid the crowded centre. This is not such a problem for those airports where a high proportion of the passengers have trip origins at the other end of the route, since they are more likely to have local destinations near the city centre or need to make onward connections.

Figure 16-2   TGV rail connection of Charles de Gaulle increased the airport catchment area (photo A. Kazda)

The rail access share depends also on the location of the station at the airport. It should be located either directly in or under the terminal building or it should have an easy connection to it, preferably dedicated rather than an urban bus that is exposed to the difficulties of the local road network.

The decision to connect the town and the airport by rail transport depends on several factors, particularly:

✈ volume of airport passengers per a year

✈ possibilities of connection to the existing transport infrastructure

but also:

✈ split of traffic between scheduled, charter, business and leisure passengers

There are several options for providing the link that should be compared. Building a special line is economically justified only for the largest airports with more than 10 million passengers a year. Provision of a rail link is now often a necessary condition of further growth of airports of this size. Most of the tracks have originally not been designed for connection to the airport and therefore the plans to do so are often compromised by the sunk investment and existing land uses. The underground or metro systems are characterised by short distances between frequent stops. Conventional heavy railway systems are characterised by higher speed, lower frequencies and longer distances between the stations. The underground is mostly suitable for the employees and for people with business at the airport. Its advantages are high frequency and low cost. The disadvantages are that it is relatively slow and not suitable for the transportation of larger pieces of baggage. For instance the journey by underground from the centre of London on the Piccadilly Line to Heathrow takes 50 minutes on average and has 18 stops. The special heavy rail Heathrow Express connects the airport with the Paddington railway station. The journey to Terminals 1,2 and 3 in the central area of the airport takes 16 minutes and to four minutes longer to Terminal 4. At present the underground to Heathrow is used approximately by 16 % of the arriving passengers and the Heathrow Express is used by 15 % of passengers Railway transport has been available at Gatwick airport ever since it opened and its high proportion in the surface transport, at present around 26 %, results from the convenient and fast connection to London and the underground network, and also to the availability of check-in at the city rail terminus at Victoria station. Munich airport has been connected by means of a rapid transit link with the city. The airport administration have set an ambitious target of achieving a 40 % proportion of railway from the total surface transport. They expect to succeed, as the rapid transit link is connected with the railway transport network.

Another question is the relation between air transport and long distance high speed railway transport. Some airport administrations consider rail to be an important competitor to air transport on short and medium distance routes. The construction of high speed lines and an increase of the quality of railway transport should, according to the opinion of this group, support this view

Others look on rail more as being an important partner of the airports, which allows a substantial reduction in the time of transportation to the airport and this increasing the catchment area. The stops of the high speed railway cannot be designed so closely as with conventional rail so they can operate only in areas with high population density. Stations for them will therefore be built only in the largest airports. At the same time these airports have the biggest problems with the number of runway slots available, particularly in the peak periods. In these cases high speed rail can really replace some regional flights and the capacity released can be used for long distance flights. But even large airports will have to contribute to the investment in the high speed railway, while it needs a convenient location close to the line of the rail track to make a direct connection to a smaller airport economically feasible. These smaller airports can be connected to the conventional network of fast trains. Those medium size airports with hub operations might absorb the regional flights displaced from the largest hubs but at the moment this does not really happen. From some smaller airports in Great Britain it is more advantageous to fly to New York through Amsterdam than through Heathrow. In some cases the airports themselves will be affected. After the Euro-tunnel opened under the English Channel, Brussels airport is more easy to access from some parts of south-eastern England than London - Heathrow.

*Figure 16-3   High speed railway station situated at the airport
(courtesy Lyon-Satolas Airport photo M. Renzi)*

Relative to rail, air transport has the advantage of flexibility, which is important particularly for small airports and small airlines. Changing routes and frequencies is much easier in air transport in railways.

A relatively new idea is to combine air and rail services. In some cases there are contracts or alliances with cross-investment, as between airlines. Since 1982 Lufthansa and Deutsche Bundesbahn have been co-operating in the operation of Lufthansa Airport Express between Frankfurt and Düsseldorf airports with stops in Bonn and Cologne. The railway carriages are in the colours of Lufthansa with the 'cabin' service of the same standard as in the air. On the train it is possible to use services such as telephones, faxes, video etc. The express train journey between

Frankfurt and Düsseldorf takes more than 2 hours, while the flight takes 50 minutes. Since 1990 another line has been opened between Frankfurt airport and the main station in Stuttgart. The Lufthansa passengers can make a seat assignment and check-in baggage in selected railway stations. The service is free of charge and the passengers can be checked in not sooner than 24 hours and not later than 20 minutes before the departure of the appropriate inter-city service. These passengers can use the special 'Rail and Fly' rate. There is also a similar co-operation between Swissair and Swiss railways.

It is obvious that a direct rail connection substantially increases the airport's attraction and, in the near future, airports will be judged on whether they have such connection or not. Those without a station directly on the airport will risk being relegated to a minor role in the system.

16.5.6 Unconventional Means of Transport

Unconventional means of transport for transportation between the town and the airport include monorail systems, magnetic cushion trains and helicopters. Hundreds of schemes were proposed in the 1970s, but very few have been implemented. The main reasons are the high costs of building the infrastructure and for the operation of the facility, the slow development of the technology and the need for planning approval in the face of objections to the land use and visual intrusion. Helicopters avoid most of the these problems but generate too much noise at the city centre heliports and under the routes, and it is difficult to find slots in the air traffic at the airport unless special non-conflicting routes can be developed. The routes tend to be at low level, so there is noise nuisance under the flight path and people living there feel that they are being spied upon on their own property. Similar objections caused the closure of the Heathrow/Gatwick transfer link.

17

ENVIRONMENTAL CONTROL

Tony Kazda, Bob Caves and Milan Kamenický

Inscription on the gate of an American airbase:

'Pardon Our Noise it is the Sound of Freedom'.

17.1 BACKGROUND

In the relationship of air transport with the environment, a number of conflicts have been identified. Air transport clearly has a negative effect on the environment. In comparison to other means of transport, however, the local environmental impact is taken only on the surroundings of the airport whereas with road or rail transport the area along the whole route is affected by noise and gaseous emissions. In compensation for the environmental disbenefits, air transport supports economic growth, as indeed do the other modes. A medium size airport provides directly several thousand jobs and indirectly tens of thousands. It is 'an economic catalyser' for its catchment area. A list of factors related to air transport that have a negative effect on the environment is given in Appendix 17-1. They have been divided into seven groups. The impact of the airport on the environment can be compared to the impact of a medium size or large industrial enterprise, with an additional special impact from the air transport operation.

One of the basic human rights is the right of an individual to a healthy environment. From this point of view air transport has been the subject of negative criticism. In the last 15 years the standard of living of those living in the developed countries has increased rapidly. The real income has increased and this opened up more free time and the wealth to use it actively. People use air transport more and more for holidays. Air transport has at the same time contributed considerably to further economic growth. As wealth increased, sensitivity to the quality of the environment grew even faster. In other words, after people had achieved their fundamental goals, they became more sensitive to their environment. The class of people with a high standard of living use air transport more and more, while at the same time more and more criticising its negative consequences. A similar development takes place also in the former Eastern Block countries with the difference that the citizens are not familiar with the appropriate legislation and the relatively small amount of air transport generates an equally small impact.

The ministers of transport of the ECAC states have concluded that the environment is one of the decisive factors in setting the future limits of airport expansion. The main target of ECAC strategy for airports is the improvement of the potential capacity of European airports and air space while preserving the standard of safety and respecting the requirements of the protection of the environment, taking into account the forecast growth of air transport in the ECAC states.

The basic strategy for dealing with the environmental issue in accordance with the ECAC conclusions can be characterised for the airports as follows:

To specify the maximum capacity in the remote future in terms of the environmental carrying capacity as well as the physical capacity of the airport site.

The impact of an airport on the environment must be based on an Environmental Impact Assessment (EIA) study. After an EIA has been performed by a specialised agency, and considered by the responsible state administrators after deliberation with other affected state administration bodies, a process of public hearings takes place with the affected communities. The length and scope of this process depends on the character of the project and character of the surrounding area, e.g. it will differ for large urban communities, compared with for nature protected areas or low quality uninhabited areas, the latter being hard to find in developed countries anywhere near a centre of population. The EIA must, above all, be based on a deep knowledge of the effected area, the character of other adjacent areas, environmental carrying capacity and restrictions, and on the knowledge of land use and compatibility planning of the site and in the communities in the surroundings of the airport.

Depending on the situation, any of the impacts could be the deciding factor in determining the outcome of an EIA, but noise is the factor that normally gets the most attention. Therefore the first part of this chapter deals with noise. It then deals with the issues of exhaust gases, ground water and bird control. The issues of ground water and soil protection are also dealt with in Chapter 9 - Aircraft Refuelling, Chapter 12 - Airport Winter Operation and Chapter 13 - Airport Emergency Services.

Appendix 17-1

FACTORS CONNECTED WITH AIR TRANSPORT THAT HAVE NEGATIVE IMPACT ON THE ENVIRONMENT

Noise of aircraft

✈ noise of aircraft in the vicinity of the airport

✈ tests of aircraft engines

✈ supersonic boom

✈ noise of aircraft on the route

Air pollution in the vicinity of airports

✈ emissions of aircraft engines

✈ emissions from the operation of vehicles on the airport

✈ emissions from transportation to and from the airport

✈ emissions from other sources on the airport (e.g. heating plant)

Factors with a global impact

✈ long distance transfer of air pollution (e.g. acid rains)

✈ greenhouse effect

✈ depletion of ozone layer

Airport construction

✈ excavation of soil

✈ soil erosion

✈ interference with ground water channels and rivers

✈ impact on flora and fauna

✈ visual intrusion

Contamination of waters and soil in the vicinity of airports

✈ imperfect treatment of waste waters

✈ leaks of oil products

✈ de-icing of airport pavements and aircraft

Waste management

✈ storing and disposal of dangerous substances used in maintenance and repairs of aircraft

✈ waste from the airport operations and from arriving aircraft

Aircraft accidents or incidents

✈ accidents or incidents of aircraft with hazardous cargo

✈ emergency procedures connected with aircraft de-fuelling

✈ other negative effects to the environment connected with an aircraft accident (e.g. fuel leakage, leakage of extinguishing substances, aircraft safety of the communities around airports)

17.2 NOISE   *(Author: Milan Kamenický, Bratislava, Slovakia)*

Noise has long been a problem around airports, Newark being closed in 1952 after demonstrations about it. Reaching an accommodation between the industry and the local communities has required progress on three fronts. Planners have needed to understand the attitude of the communities for whom they are responsible, the industry has attempted to ensure that the best feasible levels of technology are being used, and airports have had the uncomfortable task of trying to reconcile the resulting situations by mitigating the impacts.

17.2.1 Characteristics

Sound is a mechanical wave motion spread from a source. It can move through gases, liquids and elastic solids and it is connected with a transfer of energy. The speed at which a sound moves depends on the medium through which it moves.

This is the physical definition of sound. Sound can be also defined as a change of the pressure that is received by the human ears and interpreted as a sensation by the brain. The frequency could be at the minimum 20 and at the maximum 20 000 cycles per second, or Hertz (Hz). If a sound wave's frequency is higher than 20 000 Hz, it is called ultrasound. If it is lower than 20 Hz, it is called infrasound.

What is the difference between sound and noise? Clearly it cannot be the loudness only, although an aircraft can be so noisy as to injure the ear if one is too close without hearing protection. A mosquito at night-time can be more than enough to spoil a good night's sleep. Instead of considering only the physical characteristics of noise, it is possible to think of it as unwanted sound. This has certain implications.

Because noise is unwanted sound, it is closely connected with the feeling of annoyance.

Noisiness is related to the loudness of a sound, which in turn depends on when and where it occurs. Mosquitoes cause less annoyance during daytime, at least from the point of view of noise. If noise becomes loud enough, the primary concern will be the risk of hearing impairment, not the annoyance.

Noise is a specific form of pollution. It is more a sociological element than an economic one. Too loud a sound or one that is generated too often, in an improper time, in an improper place or in an improper situation is designated as noise. Aircraft noise has a disturbing effect in particular on the inhabitants in the vicinity of aerodromes and arrival and departure tracks.

Social research has shown that people differ enormously in their attitude to noise, but that their annoyance is related to both the noise intensity of a flight and the number of flights they hear. This led to the development of noise indices to represent the impact on a community, each country developing its own indices, but most containing terms reflecting both loudness and frequency.

Modern aircraft technology has considerably reduced the noise from the engine sources. The area on the ground affected by a given level of noise from modern jets is approximately 5 times smaller than that affected by the jets in 1960's. This positive trend will be in future partially diminished by overall growth of traffic and aircraft movements at major airports. However, it is important that, with the withdrawal of the noisiest aircraft from operation, the maximum noise levels for individual movements will be reduced. These peak individual noise events are one of the main reasons for complaints among the communities around airports.

Approximately only a third as many people are affected by noise at Heathrow as were affected in 1975, despite the traffic increasing by a factor of four. This is a real result, because the noise contours at Heathrow are calculated from noise measurements of actual operations rather than relying on the noise implied by certification. Yet the complaints about noise at Heathrow do not diminish, and noise is again one of the major issues in the public inquiry into the provision of a fifth terminal there.

Noise around airports can be controlled in three ways. The first way is to reduce the noise at source through improved technology and aircraft noise certification to ensure that the available technology is employed. This is not dealt with further in this book. The second way is to control the aircraft operations. This includes optimization of flight procedures, distribution of movements between runways and limiting operations by type and time of day. The process of noise monitoring in selected points of the airport vicinity is an important part of the optimization. The third way is land use and compatibility planning around the airports, particularly with regard to urbanization zones.

An approach to noise control on a system basis in all of the three areas should ensure an acceptable noise load on the inhabitants in the vicinity of aerodromes and under the arrival and departure routes.

17.2.2   Descriptors Used for Aircraft Noise Rating

Sound pressure is the basic metric for assessment of noise. Sound pressure represents time fluctuations of pressure around its static value. In a common air environments, static pressure represents barometric pressure. For common sounds, the range of acoustic pressure values is approximately from 2.10^{-5} Pa up to 20 Pa. A value of 2.10^{-5} Pa is the weakest sound to be noted by the human ear – so called threshold of audibility. That value is 5.10^9 times smaller than the normal value of barometric pressure and would be created by a deviation of the tympanum over a distance smaller than the diameter of one atom. The values of acoustic pressure of common sounds are relatively small but their range is very big $(1:10^6)$. Therefore, since sound is perceived by a human being, the Pascal is not very convenient for measuring it. Further, this is also not convenient because human perception of sound is on the basis of Webber-Fechner's physical law:

When physical intensity of the tone '**i**' of the given frequency grows in a geometric series, its subjective effect '**h**' on the human ear (tone noisiness level) increases approximately only in an arithmetic series. An approximately correct mathematical expression of tone intensity dependence on its noisiness level shall have the form:

$$i = k.a^h$$

where '**k**' and '**a**' are constants.

A logarithmic value, the so called level, is used for expressing the sound pressure and other acoustic quantities. The sound pressure level is expressed with the unit of dB, called decibel

$$L_p = 10.\log (p^2/p_o^2)$$

where the quantities with a subscript zero are reference quantities related to the threshold of audibility. The reference for sound pressure in air is 2.10^{-5} Pascals (Pa), which is equivalent in the decibel scale to 0 dB. By converting sound pressure in Pascal to decibels, a scale spanning $1:10^6$ is conveniently reduced to 1:120. Figure 17-1 indicates typical sound pressure levels for 'everyday' incidents.

Figure 17-1   The relationship between Pascals and decibels

Physically doubling the sound pressure means to increase the level by 6 dB, while a ten-fold increase means to increase it by 20 dB. On the other hand, if the sound pressure is reduced by 50 %, the level has been reduced by 6 dB, and a reduction to 1/10 corresponds to a decrease of 20 dB.

It was mentioned above that the range of normal human hearing is approximately 20 Hz to 20 kHz. However, human hearing is not linear with respect to the sound frequency. If there are two signals with a different frequency but with the same level they are perceived differently in terms of noise. In other words, sound is frequency dependent. Further, this dependence also depends on the level. This complicated phenomenon can be clarified by curves of equal loudness, called Fletcher-Munson's curves.

By reversing these curves we get the frequency dependence of sound sensitivity. On the basis of these curves, weighting curves were created and the corresponding weighting filter used for a sound level meter. The sound level meter evaluates sound in all frequencies that are perceived by human hearing. Three basic weighted curves A, B, and C, the most used being the A weighted curve and filter, were created in a way that all three chosen curve represented the same loudness 40 Phon.

Other weighted curves were also designed; of which the most used for evaluating airport noise is D. This curve was designed for evaluating sound events that occur when the airplanes overflies the measuring point. This is to capture the more disturbing part of this noise as perceived by a human being and is the Perceived Noise Level (PNdB).

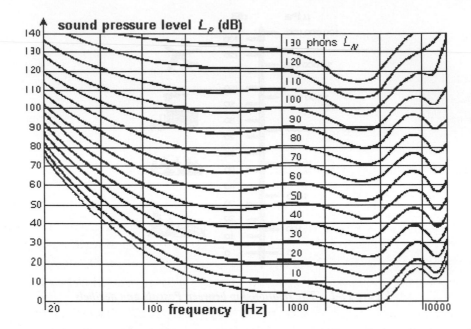

Figure 17-2   The Equal Loudness Curves

The energy content of the sound is a fundamental aspect of risk assessments of hearing impairment. The Root Mean Square (RMS) is a special kind of mathematical average value which is directly related to the energy content of the sound. It is one of the most important and most used measures, the sound pressure level being defined from an RMS value of sound pressure.

Loudness is not only a function of the frequency and the level of the sound, as has already been mentioned, but also of the sound duration. The risk of hearing impairment depends not only on the level of the sound, but also on the amount of sound energy entering the ear. For a given sound level the amount of energy entering the ear is directly proportional to the duration of exposure. Sounds of short duration are perceived to be of a lower level than steady continuous sound of the same level. To assess the potential for damage to hearing from a given noise environment, both the level and the duration must be taken into account. However, if the level is very high the duration is irrelevant. The hearing impairment will occur almost instantly with a sound pressure level over 120 dB(A). The influence of time as a factor of the noisiness effect is expressed by the quantity Equivalent Continuous Sound Level, called L_{eq}.

The equivalent continuous A-weighted sound level, L_{Aeq}, equals the constant A-weighted sound level whose acoustic energy is equivalent to the acoustic energy of a fluctuating A-weighted sound over some time interval.

The mathematical formulation is:

$$L_{Aeq} = 10 \times \log \left(\frac{1}{T} \int_0^T \left(\frac{p_{A(t)}^2}{p_0^2} \right) dt \right)$$

where:

T   duration

$p_{A(t)}$ the actual A-weighted sound pressure

p_0 the reference pressure.

This is similar to the RMS value. Although they both express an equivalent constant signal containing the same amount of energy as the actual time-varying signal itself, they cannot be substituted for one another. The L_{eq} expresses the linear energy average, while the RMS value expresses a weighted average where more recent events have more weight than older ones.

The equivalent continuous level is not the only parameter that could be used for energy content assessment. An alternative parameter is the Sound Exposure Level (L_{AE}), also called the Single Event (Sound) Exposure Level, often known for short as the SEL.

The L_{AE} is defined as the constant level acting for one second which has the same amount of acoustic energy as the original sound. The L_{AE} is the L_{eq} normalised to a one second interval. L_{AE} measurements are often used to describe the noise energy of a single event such as an aircraft fly-over. L_{eq} is normally used to integrate the succession of single events over a longer time period.

In the majority of countries, descriptors for aircraft noise issue from mathematical quantities that are defined in the standard ISO 3891 'Acoustics, Procedure for describing aircraft noise heard on the ground'. An equivalent to that standard in Great Britain is the BS 5727/1979 standard. In this standard two basic quantities are recommended for evaluation of the aircraft noise. They are an Equivalent Perceived Noise Level L_{PNeq} and an Equivalent A-weighted Sound Level L_{Aeq} for a specified period. The standard gives an exact description of the transformation of the measured quantities to those used for evaluation.

An approximate value of Effective Perceived Noise Level - $L_{EPN\ (approx.)}$ (EPNL), acquired by measuring D-weighted sound level, which expresses the total noise effect generated by an aircraft flight over the place of observance, or Sound Exposure Level L_{AE}, acquired by measuring of A-weighted sound level, is recommended as the primary quantity for determining the evaluation descriptors.

EPNL - L_{PNeq} is determined by time averaging Perceived Noise Level - L_{PN}. PNL is derived from a quantity perceived noisiness (PN).The perceived noise level is a calculated single number based on known levels of people's hearing response to fixed-wing jet aircraft flyover noise. For determining Perceived Noise Level, it is necessary to make an octave or a third-octave analysis of the sound signal.

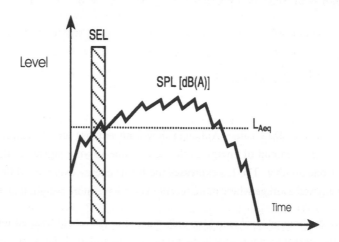

Figure 17-3   The relationship between sound pressure level (SPL), L_{Aeq} and SEL

The unit of perceived noise is PNdB. The method of computing it is described in detail in ISO 3891. Aircraft noise ranking by means of this descriptor is relatively complicated and requires a multi-spectral analyser working in real time. For normal purposes, an approximate effective perceived noise level $L_{EPN (approx)}$ is used, together with an approximate perceived noise level $L_{PN (approx)}$. The following approximate relations are valid

$$L_{PN} = L_D + 7 \text{ dB}$$

$$L_{PN} = L_A + 12 \text{ dB}.$$

At present a revision of the ISO 3981 standard is being carried out. There are expectations that aircraft noise need only use an A-weighted sound level and descriptors derived from it, such as sound exposure levels, equivalent and A maximum sound level.

However, all these metrics are only poor indicators of people's actual disturbance from noise events, this being indicated by the poor correlation of the number of complaints with the noise contours generated by the metrics. They do not adequately treat problems of noisy peak operations or noise at night, particularly in the period before waking, nor for the whining tone of the latest generation of engines, despite being corrected for the ear's sensitivity to different frequencies. There is at least a suspicion that noise complaints are actually a surrogate for fear of accidents, instanced by complaints being made even though the aircraft has been seen but not heard.

17.2.3 Evaluation of Noise in the Vicinity of Airports

Aircraft noise is evaluated in different ways in individual countries. In the majority of countries, however, only sound events that are connected with aircraft movement or other aircraft activity such as engine tests and taxiing are evaluated. Such a sound event will increase the respective acoustic indicator that describes aircraft noise against an acoustic background.

Since 1992, Great Britain has used the equivalent A sound level L_{Aeq} from sound events between 0700 and 2300 hours as the main noise evaluation tool, supplemented by a night-time L_{Aeq}, usually over the busiest three month period.

In Germany the procedures for assessing the noise load rating in the airport vicinity are elaborated in detail. They reflect experience from long-time monitoring of aircraft noise. The assessment is carried out on the basis of Act on Protection against Aircraft Noise and the DIN 45631 standard. Equivalent continuous sound level determined by sound events from air traffic in a time period of 6 months is the evaluation descriptor used.

Besides using the equivalent continuous level, the DIN standard recommends, on the basis of an analogy with ISO 3891 standard, also measuring and evaluating equivalent A-weighted sound level. Long-run equivalent continuous level is computed from measured A-weighted sound levels corrected to equivalent pressure levels over one second, the result corresponding to sound exposure level. The equivalent sound level is derived for the 6 month period for which the traffic level is highest for the airport in question. This rating level is used for comparison with noise levels from other kinds of transport and industry. In addition, the DIN standard recommends the use of the maximum A-weighted sound level measured with a time constant as an auxiliary descriptor.

In Norway two quantities are used for aircraft noise evaluation at airports. Equivalent continuous A-weighted sound level averaged over 24 hours with additional weighting for night and evening flights, similar to CNEL and L_{AeqDN} in the USA as described below, and maximum A-weighted sound level in one week. A third A-weighted sound level, maximum level, is also monitored for a period of time. Furthermore, equivalent level is averaged for the whole week with variable additional weighting of flights during Saturday and Sunday.

In France a quantity 'Indice psophique' R is used for airport noise evaluations. It is computed as a sum of noise peaks of all aircraft movements, while perceived noise level PNL is measured.

During computation and evaluation, night flights between 2200 and 0600 hours are penalised with a value of 10 dB. For computation the following relation is used:

$$R = 10.\log(\Sigma 10^{L_{PN}/10} + 10.\Sigma *LPN^{L_{PN}/10}) - 32$$

In the Netherlands a 'Rating Index' B is used. Maximum A sound levels of each sound event caused by air traffic are used. The following equation determines the evaluative index B:

$$B = 20.\log (g_i.10^{L_{Amax}/15}) - 157 \ (10)$$

$g_i = 1$ for interval of 8-18 hours; 2 for 18-19 hours; 3 for 19-20 hours; 4 for 7-8 and 20-21 hours; 6 for 21-22 hours; 8 for 6-7 and 22-23 hours, 10 for 23-6 hours.

In Switzerland the Noise and Number Index NNI(A) is used for aircraft noise rating. It is an analogy to NNI that was used in Great Britain, but it is derived from maximum A-weighted sound level measurement (with a SLOW time constant) of sound events caused by air traffic. For computation only those sound events whose maximum levels exceed 68 dB(A) between 0600 and 2200 hours are taken into consideration. Besides the averaged maximum sound level of all considered sound events, the average daily number of movements of aircraft is also taken into account in computation. NNI(A) is determined by means of the following relation:

$$NNI(A) = (L_{Amax})_{en} + 15.\log(N) - 68$$

where:

N   n/365

n   number of operations 0600-2200 hours at which L_{Amax} exceeds 68 dB(A).

In the former USSR air traffic noise is rated by maximum A-weighted sound level and equivalent A sound level.

In the basic regulation dealing with the issues of aircraft noise in the Czech Republic and Slovak Republic, equivalent A-weighted sound level and maximum A-weighted sound level (with FAST time constant) are the evaluative quantities used. The allowable limits are 65 dB (A) for L_{Aeq} + correction and a value of 80 dB(A) for L_{Amax} + correction. The correction term takes account of local conditions, according to the character of the built-up area and time of day or night. Neither limit may be exceeded. Air traffic noise evaluation is derived from a 24-hour measurement of a 'characteristic day'. Then equivalent level is determined for daytime from 0600 to 2200 hours, i.e. 16 hours and for one night hour when the one-hour equivalent noise level was the highest. These quantities are not entirely compatible with international and applied recommendations, respectively, and the evaluation period (24 hours) is relatively short. It would be more appropriate to define a longer time interval for which evaluative quantities would be averaged. Thereby a possible deviation of measured quantities from the criteria would reduce in particular for airports with non-scheduled air transport.

The existing European regulations do not solve the long-run issues of airport noise evaluation. Regulations ES 80/51 and ES 83/206 determine only maximum limits for subsonic aircraft, similar to those from aircraft certification. There are discussions about implementation of a unified evaluative quantity Weighted Equivalent Continuous Perceived Noise Level (WECPNL). In substance, it is the Equivalent Perceived Noise Level - L_{PNeq} recommended in ISO 3891, but with penalties for aircraft movements in the evening from 1900 to 2200 of 3 dB and at night from 2200 to 0600 of 10 dB. WECPNL is determined for a time range of 24 hours. That descriptor is at present recommended also by ICAO.

In the USA the issues of noise have been dealt with in detail since 1950's. Several methods have been proposed and several rating quantities have been recommended during the development. In the majority of cases they issued from the US Environmental Protection Agency (EPA) recommendations. Indicators for general evaluation of noisiness in the environment were adapted to aircraft noise evaluation. At present several indicators are in use.

The Noise Exposure Forecast (NEF) is an improved older quantity, the Composite Noise Rating (CNR), which is an analogy to NNI. NEF is derived from the effective perceived noise level L_{EPN} for individual noise events caused by aircraft movement. Noise events caused by a movement at night are multiplied by a coefficient of 16.67. For NEF = 0, a concept of 'threshold of perceived aircraft noise' was introduced. The NEF is now used to determine noise load from measurements, but was originally calculated from predictive methods.

Equivalent A-weighted sound level for Day – Night - L_{AeqDN} is another quantity used in the US. It is used for general evaluation of noise load in the environment. It is derived from a 24-hour evaluation of equivalent level. Levels for day are multiplied by a coefficient of 0.625 over 15 hours and those for night by a coefficient 0.375 over 9 hours. In addition, levels for night are penalised by 10 dB. When an evaluation is made of aircraft noise, this quantity comes from particular sound events, which are rated by A-weighted sound exposure level (L_{AE}). Simplified computation of L_{AeqDN} may be done by means of the following relation:

$$L_{AeqDN} = (L_{AeqD} - 1.8 \text{ dB}) + (L_{AeqN} + 5.2 \text{ dB})$$

Community Noise Equivalent Level – CNEL is an improved version of L_{AeqDN} used predominantly in the state of California. It is also derived from 24-hour averaging of A sound levels. For aircraft noise however it is modified relative to L_{AeqDN} so that the sound events between 1900 and 2200 are in addition penalised by 3 dB. When CNEL is determined, it is derived, as with L_{AeqDN}, from A-weighted sound exposure levels (L_{AE}).

In special cases, daytime equivalent A-weighted sound level evaluated from 0700 to 2200 hours and a nighttime one evaluated in a period of time from 2200 to 0700, are also used.

It may be stated that from the viewpoint of evaluation of the noise caused by aircraft operation, there is no consistent international regulation that would be implemented into the legislation of particular states. In every state the evaluation is made on the basis of national regulations, recommendations or laws, though some national regulations do issue from the standard ISO 3891.However, some states have calibrated their own descriptors satisfactorily against surveys of community annoyance.

17.2.4 Land Use and Compatibility Planning

In the majority of states there is an effort to provide healthy and quality environment, particularly in the residential areas and for noise sensitive activities such as schools, hospitals and recreational zones. Planning of built-up areas in the airport vicinity is an efficient instrument for preventing complaints of inhabitants in the future, as long as the plans can be reinforced. States are then able to prevent additional settlement of new inhabitants in the noise-affected zones in the airport vicinity. However, in the majority of cases, there is a question over the limitation of the owners rights to their property and therefore it is an especially sensitive area of legislation. For that reason there is an effort to solve disputes by an agreement in terms of various forms of compensation or repurchase of the estate rather than by expropriation of the estate even if the right of expropriation exists in the legal system.

In Germany the Act for Protection against Aircraft Noise determines also the limits that restrict 2 noise zones in the airport vicinity:

Zone I – equivalent noise level is higher than 75 dB(A)

Zone II - equivalent noise level is lower than 75 dB(A), but higher than 67 dB(A).

The act specifies which activities may be allowed in the respective zone, e.g. there is a strict prohibition on building schools, hospitals and new residential buildings in both of the zones.

Norwegian legislation determines four protective zones for equivalent and maximum levels.

1.zone – L_{Aeq} = 55 - 60 dB(A), $L_{Amax, DAY}$ = 85 - 95 dB(A)

2.zone - L_{Aeq} = 60-65 dB(A), $L_{Amax, DAY}$ = 95-100 dB(A)

3.zone - L_{Aeq} = 65-70 dB(A), $L_{Amax, DAY}$ = 100-105 dB(A)

4.zone - L_{Aeq} > 70 dB(A), $L_{Amax, DAY}$ > 105 dB(A)

In particular zones it is exactly determined under which conditions what structure might be realised.

Austria solves these issues by legislation as in Germany. Three zones are determined in the airport vicinity restricted by equivalent sound levels A - 70, 65 and 60 dB(A) for daytime and 60, 55 and 50 dB(A) for night operation. In addition to the Aviation Act, the ÖNORM 95021 standard also specifies the use of these zones. At present a discussion is being held about a proposal of a new 'Act for Protection against Aircraft Noise'.

In the USA, legislation determines limiting values for L_{AeqDN} as follows:

1. L_{AeqDN} lower than 65 dB(A) – normally acceptable, permitted development without restriction.

2. L_{AeqDN} higher than 65 dB(A) but lower than 75 dB(A) – normally unacceptable for all but specially designed houses.

3. L_{AeqDN} higher than 75 dB(A) – absolutely unacceptable zone, with prohibition to build any new structures other than for airport related purposes.

Similar noise criteria are defined in the UK but using L_{Aeq}. Permission for new dwellings will normally be granted below 57 dB, noise should be taken into account between 57 and 66 dB, permission would not normally be given between 66 and 72 dB, and it would normally be refused in zones with noise above 72 dB. Similar guidelines apply to night noise, but the limits are 8 or 9 dB lower.

In Slovakia and the Czech republic the land surrounding the airports is also classified into four categories of noise exposure: minimal, moderate, significant, and severe. Each category is defined by a range of one of four noise limits: maximum noise levels during day and night, and equivalent noise levels during day and night. In the areas with significant and severe noise exposure it is not possible to plan the development of urban areas or the building of noise sensitive facilities such as schools, hospitals and recreational zones.

The land use and compatibility planning is, in the long run, the most effective way to control noise around the airports. The Airports Council International (ACI) adopted a resolution in January 2000 which urges the co-ordination and unification of EU policy in this field.

17.2.5 Aircraft Noise Measurement

The measurement of air traffic noise should reflect the choice of indicator used to evaluate the noise load caused by air traffic. This will determine the type of instrument and the time for which the measurements are taken. The method and equipment used for long period monitoring will be different from those for short-term tasks such as week-long surveys. However, the results from both sets of measurements should be compatible.

In choosing the method of measurement, it is important to take into account the purpose for which the evaluation will be performed: control of the legislated noise limits, control of adherence to the chosen flight procedures, verification of the predicted data. The choice of the location of the noise meters also depends upon the purpose of measurement. For verification of prediction and control of the protective noise zones, it is necessary to chose measuring points with the lowest acoustic background possible, away from populated areas. When the purpose is to control adherence to health or annoyance limits, the measuring points are located in the inhabited zone and in the areas where noise sensitive activities are situated. In such cases it is often very difficult to identify noise that is caused solely by air traffic and special measuring apparatuses and special rating techniques are required.

17.2.5.1 Short Term Measurement

Short term measurements should serve to control adherence to the set limits for a given category of activity, to control and verify the predicted noise load in the vicinity of airports and flight corridors, and to make special measurements such as comparing noise from an aircraft with its certified values. In a short-term measurement it is necessary to observe whether individual sound events are caused by air traffic, or whether they are caused by other sound sources. The measurements should therefore always be performed with staff in attendance who identify the type of sound event as it is being measured, unless it is possible to correlate the timing of aircraft movements very accurately from radar traces. A measuring system may also by used that only provides a record of each noise event that exceeds the background noise level. The record should be qualitatively and quantitatively in such a form that will allow to the noise event to be positively linked to the cause.

The level of sound at which the recording should start may be selected. This will depend on the background level, the time for which it is exceeded, and other characteristics of the noise trace.

Short-term measurements should be performed with sound level meters of class 1 accuracy, in compliance with IEC 60651 and 60804 standards. Measuring apparatus should be able to measure simultaneously maximum A-weighted sound level and equivalent A-weighted sound level in a certain time interval. Simultaneous measurement of A-weighted sound exposure or time level should also be possible.

If a sound level meter that can measure these quantities simultaneously is not available, the sound signal may be recorded, in such a way that the time, frequency and dynamics are unchanged, in a suitable recording medium, for example by means of an analogue or digital tape recorder. Alternatively, the trace can be transformed into a digital form and saved in computer memory. Then the chosen indicators may be evaluated later with a suitable sound measuring apparatus. The recording equipment for saving the original signal should fully comply with the requirements specified by the IEC 561 standard.

If descriptors derived from perceived noise level, e.g. in aircraft certification, are used for noise evaluation, it is essential to use a multi-spectral real time analyzer with spectrum sampling of 500 ms as a minimum if detailed measurement is required. If it is only a preliminary study, A or D-weighted sound level measurements may be used for determining approximate perceived noise level $L_{PN(approx)}$, and approximate effective perceived noise level $L_{EPN(approx)}$ respectively.

Microphones for free field should be used in compliance with IEC 60651, with a cover against wind or rain. Microphones with these supplementary features should comply with the characteristics determined in IEC 561 and ISO 3891. The microphone should be located at least 1.5 m above the terrain level, or at least in a minimum of 1.2 m from the outside wall of a building if the measurement is performed in the building façade.

Before and after each series of measurements, an acoustic calibration must be performed with a microphone calibrator. When the measurement is performed during several days, the calibration must be carried out at least once a day. In the course of the measurement, temperature, air humidity and wind speed must be monitored. After measurement, the results are processed in a suitable manner into the prescribed form. They may also be used in verification of predictive methods and development of mathematical models for aircraft noise simulation.

17.2.5.2 Long-term noise monitoring

Long-run noise monitoring is performed for the purpose of a long-run control of adherence to the defined limits of noise load for a certain location, for land use and compatibility planning and for monitoring actual air traffic activity which from time to time exceeds the noise limits. In such measurement it is also necessary to monitor aircraft movement through the same airspace where the noise is being generated. Aircraft movement monitoring may be provided automatically by processing radar information. The data on the aircraft movements are matched with the noise events detected by the noise monitoring. The noise load originating only from air traffic may then be separated from other noise sources.

Unification of the results from aircraft noise monitoring and track keeping (NTK) radar data on aircraft movement may be performed in real time or additionally by appropriate data processing by computer software.

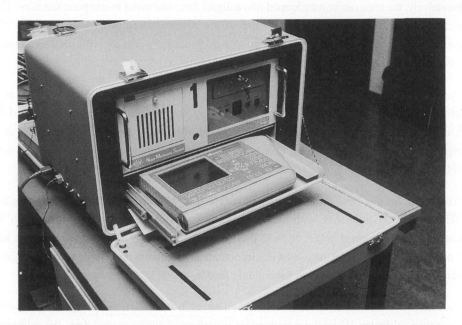

Figure 17-4   Equipment for a permanent aircraft noise monitoring with a multi-spectral analyze and audio recording from a sound events of fly-over

(Environmental Sound Analyzer ESA121, NORSONIC AS Norway, with control unit and post-processing software TOPSONIC Ltd. Germany), photo M. Kamenický

In accordance with the above mentioned, long-run monitoring of aircraft noise may be performed by means of two types of equipment:

1. Measurement is provided by means of instruments that simultaneously measure the required rating quantity (e.g. maximum and equivalent A sound levels) in the specified time interval.

2. Measurement is provided by means of instruments that simultaneously measure and record the continuous course of evaluation descriptors averaged and evaluated from a short-term interval.

Both of the systems may transmit the data from the noise monitoring units to the central control location for processing in two ways:

1. Data transmission is provided in real time, i.e. each instrument is constantly directly connected with the central measuring point and the data are continually transmitted to that point. Connection may be provided by a local network or by means of telecommunication lines. Output may be shown also on monitors in selected neighbourhood centres or airport terminals..

2. Data transmission is performed in discrete batches after certain time intervals. Data from noise measurement are saved in the local memory of the noise measuring equipment and, at the specified time, as a rule once in 24 hours, they are transmitted the central processing point. Again, a local network or the public telecommunication network may be used for that purpose.

In the majority of cases, a combined system is used in big airports. Some measuring places are continuously connected with the central control point, while the data transmission from others is intermittent. The majority of systems for aircraft noise monitoring are augmented with one or more mobile measuring stands, which are most often located in a dedicated vehicle.

Simultaneously with the sound signal measurement, it is necessary to measure also barometric pressure, air temperature and humidity, wind direction and speed. The airport operator usually performs these meteorological measurements as well as the noise monitoring.

17.2.6 Prediction of Air Traffic Noise

Mathematical models are used as the basis for computer programmes that predict the noise load in the vicinity of airports. The load may be predicted in the form of noise contours of equal levels of the chosen descriptors, and also of maximum noise levels at specific locations. Just as there is no consistent methodology for selection of the descriptors and the method of evaluating noise around airports, so there is not consistent international recommendation or directive for the prediction of aircraft noise load in the vicinity of airports.

In Germany the method of evaluating noise by prediction is determined by the Act on Protection against Aircraft Noise. Long-run equivalent A-weighted sound level is the descriptor used. The method is based on the categorisation of aircraft into noise classes. Acoustic pressure levels in octave bands, which are determined on the basis of measurement for the given aircraft categories, are used as inputs. They are integrated through the six months during which the airport traffic is the highest. Protective airport noise zones are determined by means of the computation, as described above. The computation and projection of the contours are realized by means of a polynomial formula. Noise load evaluations by prediction in the vicinity of all the larger airports, military ones included, are performed under the legislation. Each airport then has a legally valid map marked with both of the protective zones. The polynomial itself is also reported. A control of adherence to the determined noise levels in the protective zones is also defined. If an increase of more than 4 dB(A) is detected by means of long-run measurement in the outer edge of one of the zones, a new noise load evaluation by prediction should be performed in the airport.

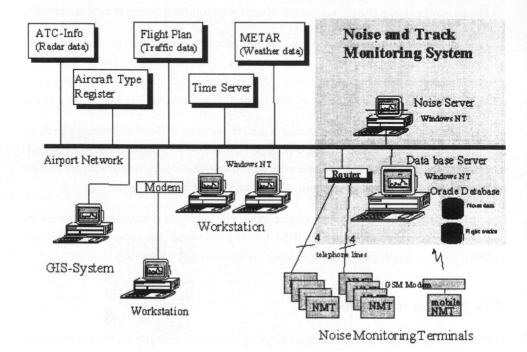

Figure 17-5   Aircraft Noise Monitoring Network
Source: TOPSONIC, Ltd. Germany

In the USA several methods for computing were elaborated in the past. The main descriptors used are L_{AeqDN}, NEF and CNEL. The US Government has financed though the Federal Aviation Administration the Integrated Noise Model (INM), that allows the noise load to be determined consistently in the vicinity of any airport. The programme allows the majority of the noise load descriptors to be calculated. A similar programme called NOISE MAP was developed also by the USAF for their own requirements.

In all these models, the path of an aircraft is approximated by means of a series of straight flight segments. In each segment, time-integral computation of target descriptors, which are computed from sound exposure level or perceived noise level of individual sound event, occurs. The maximum A-weighted sound level contours and maximum perceived tone-corrected noise level (L_{PNTM}), are computed directly from these descriptors.

The computation at individual points of the grid map is performed separately for each independent aeroplane that is considered in the model. Noise characteristics of individual types of aircraft acquired from aircraft certification or from special measurement may be used as input data for computation. The data collected during certification are incorporated into the INM. However, these data are not part of the public dissemination of certification data, the latter only covering the noise results for the three measuring points specified in Annex 16. In the event that data on maximum A-weighted sound levels or maximum perceived tone-corrected noise level are not available for some types; these will be computed by means of an empirical formula based on linear regression. In the computation, the effects of flight characteristics, procedures and aircraft configurations during take-off and approach are taken into consideration.

Recently there has been a considerable effort to unify the predictive methods used for determining aircraft noise at airports. Unfortunately no useful conclusion was reached. ICAO recommend the method for computing noise given in ANNEX 16. The use of this method allows comparison and evaluation of airports on an international basis.

17.2.7 Airport Noise Mitigation and Noise Abatement Procedures

Airports need to adopt noise mitigation measures in order to pacify their local communities or simply to behave in an environmentally responsible way. The airport operators use measures in addition to noise abatement approach and departure procedures. They include displaced thresholds, preferential use of runways which demand more crosswind and tailwind movements than some airlines are happy with, restrictions on the use of reverse thrust, limits on daily movements, curfews and night quotas.

The takeoff noise should mainly be a problem for the heavy four-engined aircraft, since the performance of the twin-engined aircraft could allow them to be at a comfortable altitude before crossing the airport boundary. This is compromised by the use of reduced thrust and by intersection takeoffs. These are two examples of the way today's operating procedures differ from those used in certification. A further example is the height at which the thrust is cut back, which is more important for the noise-critical large four-engined aircraft. In certification, this is done to minimise noise at the certification point 6.5 km from start of role, whereas it is normally controlled by specific noise abatement procedures, instructions in the operating manuals or operators' preferences. If this class of operations is to operate satisfactorily, it is essential that the procedures are chosen to minimise noise at the most sensitive locations.

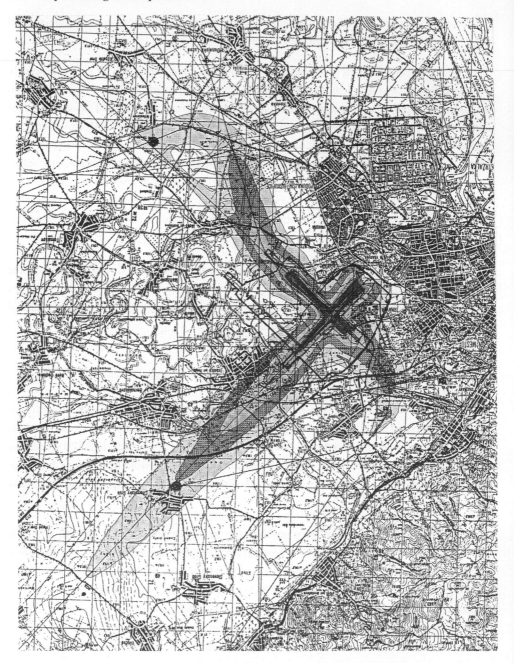

Figure 17-6   Noise contours of equivalent A-weighted sound level for day - prediction of noise rating around Bratislava airport, Slovakia, courtesy Bratislava airport

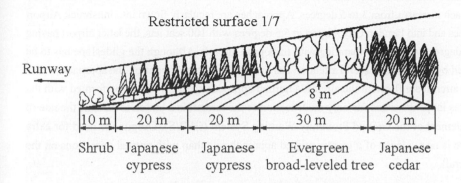

Figure 17-7 The noise protection wall
Source: Airport Planning Manual, Part 2, Land Use and Environmental Control

The increasing use of high bypass twin-engined aircraft has transferred attention from the takeoff noise to the final phases of approach, where the dominant noise affecting L_{eq} contours comes from either the front fan or from the airframe. This phase is difficult to improve, because the aircraft needs to be stabilised on a standard three degree glideslope. There is, however, also increasing concern about noise further out under the approach path. Many complaints coming from these areas in the early morning.

The noise under the early stages of approach could be mitigated by changes to the flight procedures. The two segment approach produces the expected large benefits between 20 and 10 nm, a slight disbenefit at 7 nm being due to the relatively early selection of landing flaps. Both the Continuous Descent and the Decelerating Approaches offer a good proportion of the two segment approach benefits, the latter being the more preferable, the benefits all coming before 8 nm from touchdown.

Closer to the airport, where noise levels would be expected to cause more annoyance, one of the main measures used by national regulators and airport companies in attempting to limit arrival noise is the stipulation of minimum altitudes for joining the Instrument Landing System (ILS). At Heathrow this altitude is 2500 ft during the daytime and 3000 ft between 2300 and 0700 hours. At Birmingham, Gatwick and Stansted the stipulated minimum is 2000 ft.

If the approach can be flown at a higher angle all the way to the flare before touchdown, the noise benefit seen in the two segment approach can be extended through the areas of greatest noise. The potential benefits would come from the greater distance between the aircraft and the ground, and also the lower thrust levels required for greater descent angles in the landing configuration. Airbus has estimated that the A310's 80dB(A) contour area decreases from 1.45 km^2 to 0.51 km^2 when the approach increases from 3 to 5 degrees. Approaches are regularly flown into Innsbruck Airport at 4 degrees and into London City Airport at 5.5 degrees with 100 seat jets, the latter airport having used 7.5 degrees while only operating with turboprop aircraft. Although the glideslope has to be intercepted by 3000 ft and careful attention has to be given to speed control, the approaches with BAe 146 aircraft are flown at the aircraft's normal approach speed, it can be performed with the dive brakes inoperative and the only modification to the aircraft is a button to turn off the rate of descent element of the Ground Proximity Warning System (GPWS). Despite the need for extra care, there is no evidence of a greater missed approach rate than with normal operations on the same aircraft.

Up to 4 degrees, it is assumed that a normal jet aircraft can be used at the normal approach speeds but lower thrust settings. The normal aircraft could still be used at higher angles, particularly if the landing weight could be decreased, but this is unlikely on a regular basis. For angles greater than 4 degrees, it is therefore necessary to redesign the aircraft to reduce landing speed for a given weight.A problem with steeper approaches at the moment is therefore that not all aircraft could fly them, so it is necessary to segregate the traffic in some way. This may be by displaced thresholds or approaches to new short parallel runways, which would give fewer ATC problems.

17.3 CONTROL OF GASEOUS EMISSIONS

Air transport consumes only 5 % of the world consumption of oil products. Every year there is an improvement in the efficiency of energy utilisation of 3 to 4 %. Jet aircraft account for approximately 2 to 3 % of the world-wide production of Nitrous Oxides (NO_x) and Carbon dioxide (CO_2.). Yet the turbo-jet engines built after 1982 produce 85% less unburned hydrocarbons (HC) than the engines built in the seventies. Carbon Monoxide (CO) exhaust gases decreased by 70 % in the same period. Air transport contributes approximately 1 % to global warming as a consequence of CO_2 emissions.

The procedures of emission certification of aircraft engines, limits of harmful substances and specifications of metering devices are included in Annex 16, Part II. However it is not the objective of this book to deal with the reduction of harmful substances at their source. There are operational measures that decrease the volumes of products of combustion at airports, including optimising the movement of aircraft on the apron and taxiways, operational towing of aircraft to the holding positions for take-off and taxiing after landing with minimum necessary number of engines running.

Reduction of emissions in airport operation can often be achieved by the same procedures which are used for decreasing noise pollution and fuel consumption. During the take-off and climb it is possible to achieve a reduction of emissions by selecting the optimum aircraft configuration and engine setting. Flight paths can be shortened during approach and landing. Descent with higher approach speed and with lower flaps settings reduces aerodynamic drag, so requiring less thrust and decreasing noise. It may also result in a reduction of exhaust products, depending on the product and how it varies with engine power. The amount of CO and HC produced per kilogram of fuel burnt reduce with increase of power, while NOx increases.

From Manchester 1999 Environmental Report - Air quality objectives:

Nitrogen Oxide
+ Annual hourly mean, 21 parts per billion (13.3)
+ Maximum hourly mean, 150 parts per billion (63)
+ Sulphur Dioxide, Maximum 15 minute mean, 100 parts per billion (83)
+ Oxone, Running 8 hour mean as 97th percentile, 50 parts per billion

Waste management:
+ 8686 tonnes generated, 500 tonnes (5.8%) recycled, target of 15% by 2005

Aircraft generally contribute much less to air pollution in the vicinity of the airport than the power generation at the airport and particularly surface vehicles on the airside and access transport. It is necessary to pay the utmost attention to the design of the system of surface transport to reduce the pollution as much as possible, as discussed in Chapter 16. The FAA's Emissions and Dispersion Modelling System (EDMS) can be used to assess and model the quantity of gaseous emissions produced in the vicinity of airports. Future levels can also be estimated if the characteristics of combustion and energy use can be predicted.

It can be assumed that the worst contamination of ambient air on the apron occurs in the operational peaks in the summer period. In the future it will be necessary to introduce a higher proportion of electrical ramp handling equipment. At the same time it will be necessary to decrease the use of aircraft auxiliary power units (APU) by using the central supply of electrical power, air conditioning and other media directly to the aircraft by lines located on passenger loading bridges or built into the apron.

The experience of most developed countries shows that the quality of ambient air in most cases is lower on airports than in the cities, the exception being by the drop-off and pick-up kerbs in front of the terminal.

17.4 PROTECTION OF WATER SOURCES

Construction of the airport and in particular construction of the runway can disturb the ground water system not only in the area of the airport but also in the wider surroundings of the airport. Rain water drains quickly from the paved areas, which can cause flooding in the water courses that are fed by the rain drains. Flood waters are therefore usually fed into retention tanks.

In the new Terminal 2 of Frankfurt Airport the rainwater is being used for services. Rain water is used for cooling of air conditioning, in fountains and in 800 lavatories, which consume approximately 8000 m^3 of rainwater a year. Before use the water is chemically treated and cleaned in sand filters.

Most countries have standards that set the water quality. However, the limits set by airports are usually more stringent. For example in 1995 Manchester Airport imposed following max. concentrations limits: 10 mg/l BOD, 50 mg/l SS, 2mg/l ammoniacal nitrogen, no visible oils.

Before entering the sewer the waste waters must be properly treated. This treatment of the waste run-off from the operations of washing of vehicles and aircraft, from the accumulator room, the boiler house, the degreasing workshop, the kitchen and dining hall must meet other health and safety requirements regarding the contents of heavy metals, chlorinated hydrocarbons and sedimentation substances.

Sewage water from the airport is cleaned in normal water treatment plants. Drainage form the movement areas of the airport require a special treatment due to the oil products or de-icing substances used in the winter. Rain water from the paved areas, in particular from the apron, can be cleaned in a special treatment plant on the airport, with separation of oil products, or the collector can be connected to the local municipal treatment plant. Special procedures have to be prepared and put into operation in the case of a large oil spillage. In addition to the measures to cope with an oil spillage, the contingency plan should contain the following:

- ✈ procedure for removing spilt fuel
- ✈ procedure for closing or sealing the sewage system
- ✈ procedure for reporting the accident
- ✈ prohibition of routine and planned maintenance on the areas not equipped with a trap for oil products
- ✈ minimising the washing of aircraft on the apron.

Particularly dangerous areas, e.g. fuel stores, hangars and maintenance workshops, have to be equipped with traps for oil products and inspected once a month.

At the same time it is necessary to pay attention to possible contamination during fire fighting training. These training activities should be permitted only within a specially dedicated area.

17.5 LANDSCAPING

Air transport needs less than 8 % of the land needed by rail transport and less than 1 % of that needed by road transport. In relation to the number of transported passengers, the land is used 5 times more efficiently by air transport than by railways and 6 times more efficiently than by roads. None-the-less, airports can cause considerable visual intrusion. They are best placed on elevated ground in the interests of clear flight paths, so are likely to alter the skyline with levelled ground and tall buildings, particularly the control tower. At night the approach lights and the apron floodlights are a distraction, the more so because airports are normally sited in open countryside.

Planning the development of an airport has to consider not only the airport site but also its surroundings. The planning must be integrated into the urban planning of the region. In addition to other impacts, the impact of the airport on the appearance of the landscape must be assessed. It is particularly difficult to situate or hide large airport facilities like hangers and terminals so that they do not disturb the appearance of the landscape. The new terminal at Bristol in the UK was limited to one floor on the airside face to minimise its impact on the skyline. Changing the landscape in the vicinity of the airport, such as tree planting, can be used to minimise the negative impacts of the visual intrusion. Landscaping by the creation of planted buffer zones can also help to control air pollution and minimise the perception of noise. However, the types and composition of the trees must comply with recommendations for discouraging birds. In this connection, it is necessary to pay attention to the treatment of waste from the airport. Earthworks can be built as acoustic barriers but can reduce the view and create their own visual intrusion.

Landscaping is also important in managing the habitats and hence the species of flora and fauna around the airport. Careful planning can avoid habitats being isolated by leaving movement corridors, provide habitats for endangered and protected species, and discourage heavier birds.

Together with improvement of the environment on the airport, airports are increasingly utilising their land as recreation zones not only for the visitors to the airport but also for its employees and inhabitants of adjacent cities. Sports grounds, golf courses, areas for water sports etc. have been located at airports. These activities, together with other airport services, are a significant source of airport revenues.

17.6 WASTE MANAGEMENT

An airport with a capacity of approximately 5 million passengers a year can generate as much waste as a small town. Airports generate between 0.5 and 1.0 tonnes per annum per 1000 passengers. To minimise the negative impact on the environment in connection with waste dumping, it is necessary to look for ways to decrease the quantity of waste particularly by recycling, reuse and introduction of wasteless technologies. In developed countries the concepts of separation and recycling are supported by legislative measures. These measures include tax relief and direct state subsidies. This will further be accelerated by the fact that in all countries there will be a gradual increase of prices for dumping of waste.

Airport waste contains a high proportion of substances and raw materials which really need to be sorted out and recycled. The sorting of waste must be considered not only as a measure for protecting the environment but also as a financial asset.

One of the main problems in the countries of the former Eastern Block and in the third world is a low social conscience. Economic growth is felt to be more important than the problems of waste management in these countries. It is therefore important to pay increased attention to the education of the airport employees in waste management, in particular in the key workplaces which are the main generators of waste.

The basis of waste management is ascertaining the composition of the waste. The information can be obtained from the waste disposal company, complemented by an analysis of the waste into individual fractions.

Depending on the airport size and the location and type of the individual workplaces, the sorting of waste can be organised as centralised and decentralised or as a combination of both procedures. The centralised collection and sorting of waste requires high investment and personnel costs, but provides better tidiness and separation of individual fractions and therefore also better opportunities for negotiations with consumers of classified waste. Decentralised collection can allow certain kinds of waste, such as paper and carton, to be separated at each location. Biodegradable organic materials, which can be composted, should be separated on the airport and this can decrease the total quantity of waste. A difficulty is that organic waste from aircraft must be considered as infectious material. Therefore in some airports it is sterilised before further processing.

The airport company can offer the waste management system to other firms in the airport, because sorting out and recycling of waste are ineffective for small enterprises. For example, Amsterdam Schiphol's administration offers the possibility of separation and processing of waste to individual organisations at the airport. The total costs for such collection are about 25 % lower compared with the classical collection process, where most of the waste is burnt. Waste can be solid or liquid, organic or inorganic, non-hazardous or hazardous. As many as 10 kinds of materials may be separated, but most organisations use three to four categories of separated materials. The main separation parameters are mass and toxicity. In the future, the containers will be equipped with electrical chips, which will not only automate invoicing but in case of contamination of separated material by other waste, e.g. paper by metal, plastic etc., they will at the same time be able to trace the transgressor. However, the best waste management philosophy is not to let the waste be generated in the first place. It may be that the purchasing strategy should be reviewed to avoid unnecessary packaging or to emphasise materials with a low toxicity.

17.7 BIRD CONTROL

17.7.1 Introduction

The number of aircraft increases continuously. At the same time their speed increases and a great number of flights is performed in low altitudes by general aviation, military aviation and by commercial aircraft on landing and takeoff. Most birds also fly at less than 300 m above the ground. It is therefore inevitable that bird strikes are a continuing serious hazard for aircraft. At the very beginning of aviation, the low speed of flight made it possible for the birds to avoid a collision and the force of any impact was small. In most cases a collision resulted in minor damage to wind shields and leading edges of wings or the fuselage. The probability of collision was also small because of the small number of aircraft. Despite this, the test pilot Carl Rogers died after a collision aircraft between his aircraft and a bird in California in June 1912. His Wright Flyer hit a seagull, which jammed the steering mechanism and the aircraft and pilot fell into the sea.

Most of the birds quickly became used to the noise and speed of propeller aircraft and learnt to stay away from the dangerous airspace in the vicinity of the airport. The situation changed in the fifties after turbo-prop aircraft and jet aircraft were introduced. This increased the speed of flight, the aircraft's acceleration and their dimensions. Therefore it was more difficult for the birds to avoid the flying aircraft and the impact force of a collision was greatly increased. The rate of collisions was also increased by the large quantity of air sucked into the jet engine and large dimensions of the engine intake. The jet engine also proved to be less resistant than piston engines to collision with the birds. Moreover, modern aircraft are even more quiet and therefore it is easier for them to escape the attention of the birds.

The cost to an airline as a consequence of damage to an aircraft after a bird strike can reach millions of dollars. It is not only the price for the repairs of the aircraft but more the costs in connection with taking the aircraft out of service, redirecting passengers and freight to other flights and the costs for accommodation and board of the passengers. Delay of the aircraft or its temporary withdrawal from service has a domino effect on the time table of airline and all connecting flights. It can cause a substantial increase in operational costs and blemishes the good reputation of the airline. On the other hand, the costs to airports in preventing collision of birds with aircraft is also not negligible. On New York's J.F. Kennedy airport they exceed half million dollars a year, which is approximately the price of repairing two engines on a B 747 aircraft.

The size of the bird strike problem is different at each airport. In order to prevent bird strikes effectively in the vicinity of airports it is necessary to have an in-depth understanding of the customs and behaviour of the birds. It is therefore important for the airport administration to co-operate with ornithologists.

17.7.2 Bird Strike Statistics

In November 1979 ICAO requested all member states to report all collisions of aircraft with birds. An international system of collection and evaluation of data on collisions of aircraft with birds was set up, called the ICAO Bird Strike Information System (IBIS). A bird strike reporting form was sent to the member states. In order to maintain the continuity of data, this form is still used. Later a supplementary questionnaire was prepared for airlines, in which the costs in connection with the bird strike and detailed information on the damage to the engines should be entered. Analysis of approximately 35 000 bird strikes hitherto reported indicates that:

✈ the number of incidents with significant damage to aircraft (1924 damaged aircraft) represents approximately 5 % of the total number of bird strikes

✈ 69% of bird strikes take place in daylight, 15% take place at night and the rest of bird strikes take place at sunrise and sunset

✈ 65 % of damaged aircraft with a weight above 27 000 kg have engine intake damage

✈ 29 % bird strikes take place during the approach and 25 % during the take-off run

✈ 51 % of bird strikes take place below 30 m above the terrain

✈ in 91 % of cases the pilots were not warned of the appearance of the birds.

Collection of data on birds strikes by ICAO is only one part of the total collection and evaluation of data necessary for the functioning of an efficient airport system of bird management. In fact the data from the IBIS system are only of small significance for any specific airport. It is necessary to complement these data by observations of the appearance of birds, maintenance reports, observations of pilots and air controllers and ornithologists on the airport or in its vicinity. For example, the pilot's report on the 'close approximation' of the aircraft with birds in the vicinity of the airport is just as important as the bird strike report, showing, as it does, that birds are a threat. The airport report should not be focussed only on whether there is a bird strike but also whether and where the birds appear in the vicinity of the airport. Only after all data had been processed and assessed it is possible to identify the scope of the problem and to propose and introduce an effective programme, the objectives of which are the decrease of the number of birds in the vicinity of the airport and the avoidance of bird strikes. The reporting procedure has to be co-ordinated by one person and all airport employees must be acquainted with it.

The aircraft can be damaged by any kind or size of bird. The extent of the damage by the bird depends on the weight of the bird, speed of bird strike and on the strength of the aircraft structure. During the impact, the speed is changed within a very short period. The change of momentum of the body is proportional to the change of the speed and the change of momentum of the body equals the impulse force.

In mild climates there are indigenous, transient and migratory birds.

+ Indigenous birds live all year in the same small area.

+ The transient birds appear in a certain area only at nesting time. Then they move to other localities or roam over a vast area

+ During the year the migratory birds fly long distances. They have a specific direction and target, to which they fly by permanent 'flight routes'. Their nesting and hibernation places are very distant from each other.

The birds have their natural daily cycle, just as humans do. The beginning of the activity, as the birds wake up, cannot be clearly determined. Some kinds of birds wake while it is still dark, others only after sunrise. The beginning of the activity is influenced by meteorological conditions. If the sky is overcast, the birds wake later.

The total number of birds changes in the course of the year. Most kinds are represented in central Europe during the spring and autumn migrations. In this period the birds are also most mobile. The spring passage is the shortest, while the autumn migration has been spread out over a longer period. The flight altitude is significantly affected by meteorological conditions. Smaller birds particularly do not want to lose visual contact with the ground. The average height of migratory flight is between 100 and 300 m.

The population of birds never spreads out evenly over a large area. The ecological conditions ensure that the birds are located in some 'islands'. As a rule, several species of birds appear in any specific environment, or habitat. The habitats can be:

✈ original and natural

✈ affected by human activities.

An airport's development normally means removing vegetation, trees and shrubs. This can cause the disturbance of the natural balance of habitats, and the eradication of existing plants and animals. Therefore it is necessary to consider the landscaping of the countryside so that it does not seriously disturb the ecological stability of the close vicinity of the airport, e.g. the ground water system. On the other hand the safety of air operations must not be endangered by an increase in birds, fog, dust, etc.

17.7.3 Change of the Habitat

The reduction of the number of birds in the vicinity of the airport can be achieved by changing the habitat so that it is unattractive for the birds. It is a system measure, which is often expensive but if the measures are considered during planning of the airport, the increase of costs is insignificant in comparison with the resulting benefits. System solutions connected with the change of habitat should represent perhaps 70 % of the total bird control effort.

The preconditions for birds to inhabit the airport are in particular:

✈ feeding

✈ safe nesting

✈ safe resting

✈ en-route passage above the airport.

The possibility of feeding is probably the most important attraction. Waste dumps are very attractive for the birds. The construction of new dump should not be situated closer than 13 km from the airport. The only effective method which can prevent birds feeding in the waste dump is to cover the dump by nets. It often happens that the visitors to the airport feed the birds. Earthworms are a delicacy for the birds, and there are many of them as well as other insects in the grass and on the paved surfaces after rain. Swarms of insects form over runways that have been heated by the sun, and they attract the birds to this worst possible location. In the vicinity of the airport, birds of prey can catch mice, moles and other small rodents. Birds are also encouraged to find food on small ponds with an abundance of small fish, frogs, larvae and water plants. Significant food for the birds are also various berries, seeds and agricultural plants.

Because of the great variety of various kinds of bird food, it is practically impossible to remove the sources from the airport. For example, during mowing of grass and hay collection, the birds collect the insects behind the machines. In addition to the elimination of natural and artificial sources of food it is possible to use chemicals to reduce the population of insects and rodents. Small water courses and swamps must be dried and backfilled, as in the vicinity of the Charles de Gaulle airport, or covered by nets, as at Bordeaux/Merignac or Marseilles/Marignane airports, in order to prevent the birds from accessing food. Attention has also to be paid to the system of tilling soil which has to be agreed by the airport administration with the owners of surrounding land. Hangars and other premises on the airport can create safe nests. The birds nest directly on and in the buildings, bushes and in other greenery, and sometimes also in the parked aircraft. On the aircraft themselves they nest most frequently in the gaps between the aerodynamic controls and the fixed surfaces, and in the engine nacelles. The nest can be built very quickly, within several hours. The nest can jam the controls, as can objects that are collected by the birds and stored there by magpies, jackdaws, crows, etc.

In addition to modifying the construction of hangars and buildings in order to prevent penetration of birds into the buildings, it is necessary to eliminate other suitable places for nesting on the airport. It is advantageous to keep a forest a safe distance from the runway, as it will be an attractive place for birds to live.

In connection with the development of Indianapolis International airport, a nesting place of swamp bats, which are listed as an endangered species, was destroyed. The administration of the airport decided to build a new suitable nesting place for the bats. They bought approximately 550 acres of forest and agricultural land along a nearby brook, which will be rebuilt as a swamp. 3000 nest boxes for the bats will be located in the forest. The total price of the forest is 3,3 million USD.

The wide and level area of an airport gives the birds a secure feeling. They cannot be attacked by any enemy without being observed. Seagulls feel so secure that sometimes they bring their food to the airport specifically to eat it there. They make the runway surface dirty, and if they are scared the whole flock rises and makes a wall which cannot be avoided by an aircraft.

Special attention has to be paid to the maintenance of the runway grass strip and other areas covered by grass. It is best if the grass height is kept around 20 cm or more. Only a few species of birds like pheasants and partridges prefer long grass, whereas most birds rest on short grass or they search for food in it. If the grass is higher than 20 cm, the view of the birds is limited and they cannot move easily in the grass. In order to avoid frequent grass cutting and also to make it less attractive, there are special grass mixes, which grow up to the height of around 20 cm and in addition to this, they contain particularly sharp and thorny blades.

Birds get used to attempts to disturb them, and take refuge in the areas adjacent to the airport. Their reactions are unpredictable and are also a serious danger to air operations. Indigenous birds, those that live in the vicinity of the airport the whole year, are familiar with the air traffic operation and can avoid the dangerous places, except when young or below average intelligence. A bigger problem is represented by 'strangers' i.e. those birds which fly into the airport from another locality.

If the airport is located on a migratory route, it may be used by the birds as a resting place. A great number of birds can flock together in this way. The only effective counter measure is not to build the airport in this locality. Information on the movement of birds in the vicinity of the airport are published in the air information manual (AIP).

17.7.4 Bird Scaring

It is not likely that the changes in habitat will completely succeed in removing the birds from the vicinity of the airport. In fact, at any time a flock can fly onto the airport and sit on the runway. The sporadic appearance of birds at the airport must then be controlled by scaring them off, particularly when there are air traffic operations. The flock of birds must be scared and dispersed as soon as possible, because a flock of birds on the ground attracts other birds of the same kind and the flock gets bigger. Also, it is easier to disperse a small flock than a big one. The easiest way is to scare the birds before they settle down on the airport.

Bird scaring methods vary in their effectiveness depending on the situation. Some techniques can be used only sparingly, because the birds get used to them. It is often advantageous to combine the techniques or to change the intensity of use. Continual scaring can, none-the-less, significantly reduce the number of birds in the airport. The bird scaring schedule must be set in advance in order to avoid repetition. Only in this way is it possible to ensure that the birds do not get used to the scaring procedure. The birds react immediately to some methods and fly away, at other times they only fly away after some delay. Different species of birds react differently to various techniques.

The success of each technique is apparent immediately after it has been used. Scaring techniques can be divided into:

✈ sound

✈ visual

✈ mechanical

✈ catching of birds

→ falconry

→ chemical.

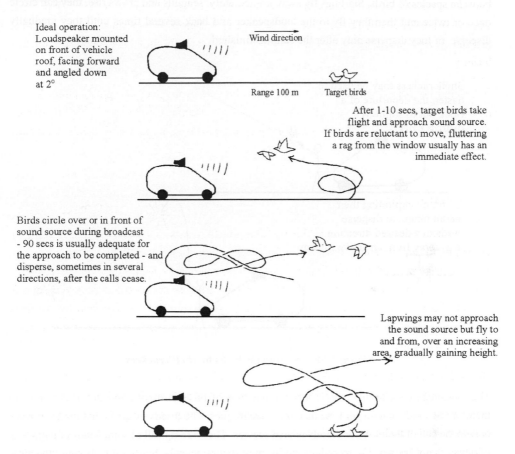

Ideal operation:
Loudspeaker mounted
on front of vehicle
roof, facing forward
and angled down
at 2°

Wind direction

Range 100 m   Target birds

After 1-10 secs, target birds take
flight and approach sound source.
If birds are reluctant to move, fluttering
a rag from the window usually has an
immediate effect.

Birds circle over or in front of
sound source during broadcast
- 90 secs is usually adequate for
the approach to be completed - and
disperse, sometimes in several
directions, after the calls cease.

Lapwings may not approach
the sound source but fly to
and from, over an increasing
area, gradually gaining height.

Figure 17-8   The bio-acoustic method of bird scaring

Among the sound techniques are:

→ bio-acoustic techniques

→ artificial sounds

→ pyrotechnical means

→ gas guns.

The bio-acoustic technique is a very successful method of bird scaring. It is based on reproduction
of the records of distress calls, warning calls, sounds of birds of prey, or command sounds which

are made by the flock of birds when taking off. The most authentic sounds are records obtained directly in the natural conditions of birds reacting to danger. The reaction to warning calls differ between species of birds. Starlings fly away immediately, seagulls and crows rise, they can circle once or twice and then they fly to the loudspeaker and back several times until they gradually disperse, or they disperse only after the call has finished.

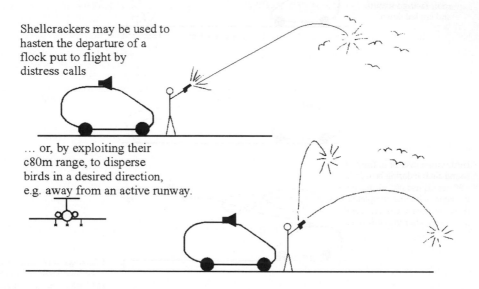

Shellcrackers may be used to hasten the departure of a flock put to flight by distress calls

... or, by exploiting their c80m range, to disperse birds in a desired direction, e.g. away from an active runway.

Figure 17-9 Scaring the birds *by shell crackers*

The advantage of warning calls is that the reaction of birds is inborn and this technique can therefore be used for a longer time than other techniques. The disadvantage is that the birds react only to the call of their own or closely related species. The warning call of some kinds of birds, e.g. of doves, is not known. The recording and its reproduction must be high quality, because the birds are capable of recognising that this is not the real call. It takes time to scar birds. Therefore it must not be used immediately before the take-off or landing of the aircraft, because the birds could just fly into its path. Fixed or mobile facilities can be used for scaring. The advantage of mobile facilities is that they are cheaper and more effective than fixed systems, and the birds do not get used to them so easily. It is advantageous to combine the bio-acoustic method, e.g. with shooting a 'very' pistol, but only **after** the call has stopped.

Some airports have successfully tested scaring birds by artificial sounds of various frequencies. In most cases this technique used stationary loudspeakers installed on the approach and along the runway. In 1982 this equipment was installed on Charles de Gaulle airport.

After pyrotechnical methods have been used, the birds rise and usually fly away faster than after the use of bio-acoustic techniques. Special Very pistol shells are used for scaring the birds. The first effect is the blast from the gun, then during the flight the shell glitters or smokes and at the end of the trajectory it explodes for a second time. The disadvantage is that the birds will be driven away only about 80 m and after a while they will return. It is better to work in two directions at the same time with weapons for and to herd the birds gradually from the danger area.

The gas gun automatic noise scaring device is a modified acetylene lamp with automatic fuse. The disadvantage is that explosions are periodically repeated. The birds find out very quickly that the gas gun is not a real danger for them and take little notice of it, unless this technique is not sporadically complemented by e.g. live ammunition. Gas guns are therefore less used these days.

Among visual techniques of scaring are:

✈ driving off

✈ scarecrows, ribbons or flags

✈ lights.

The most effective is if the birds are scared by people who face the birds and move their hands approximately 24 times a minute. They should be standing above the level of the birds' horizon (Figure 17-10). The birds consider them to be a large attacking predator. Visual techniques of scaring can be used in noise sensitive areas or when the use of pyrotechnics could cause a fire risk.

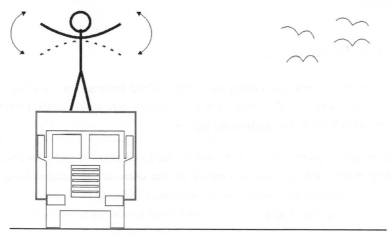

Figure 17-10 Scaring the birds by moving hands

Static scarecrows or other mechanical means for scaring birds are not effective. Flashing lights, e.g. on the roof of a car, are used in combination with other methods, e.g. shooting. The birds gradually develop a conditioned reflex to this. A variation is to drive a car into the flock of birds while blowing the horn.

Among the mechanical techniques are nets or thin wires, which prevent the birds gaining access to food. It also includes using plastic or metal soffits on buildings, which are smooth and thus make building nests more difficult.

Endangered species of birds can be caught and relocated into their new locality. Bird catching is highly qualified and time consuming work, seldom done by airport employees.

The use of falconry, scaring birds by their natural enemies, namely birds of prey, is only seldom practised. The reasons are high acquisition and operational costs. In addition to the trained bird of prey, an all terrain vehicle, space for locating of the birds and training premises, aviaries, guns, hunting dog etc. also have to be available. In some countries falconry has been forbidden, because the only suitable predators are protected as endangered species. There is also the issue of the legal responsibility of the airport administrator for the damage in the event of a collision between an aircraft and the trained predator. The bird of prey cannot be used in bad weather and it is not effective for all species of birds.

Use of chemicals is strictly limited in many countries. These are mostly used for reduce the population of insects or rodents, which could serve as feed for the birds. In order to prevent nesting or resting of birds in high inaccessible places like flood-lighting masts, special pastes are used which repel the birds.

17.7.5 Ornithological Protection Zones

In some states ornithological protection zones are defined around airports. The objective of the declaration of ornithological protection zones is the limitation of bird strikes in the vicinity of airports, where the probability of bird strike is the highest.

In the inner ornithological protection zones, no dumps, stacks or silage must be located. The method of tilling agricultural soil will be agreed by the users of agricultural land with representatives of the airport administration. In some cases, areas belonging to the airport are leased out for agriculture. In these cases it is necessary to agree planting of suitable products.

Outer ornithological protection zones surround the inner ornithological protection zones. It is possible to establish agricultural facilities, such as chicken and pheasant farms, cowsheds, centres for waste collection and processing plants, lakes and other structures and facilities in any significant quantity in this zone only with the approval of the Aviation Authority.

18

PREDICTING DEMAND

Bob Caves

Motto: The only certainty about the future is that it will not be as forecast.

18.1 INTRODUCTION

The process of airport planning requires a large number of forecasts to be made. Most attention is given to the prediction of passenger and freight traffic. It is, however, also necessary to attempt to predict likely changes in aircraft and airport technology, productivity in passenger and freight handling, choice of access modes, number of airport workers and the number of 'meeters and greeters' per passenger, together with airport cost and revenue per passenger. Also the airport planner is not alone within the air transport system in needing forecasts. The airlines, the manufacturers, the sub-system suppliers and the national transport planning authorities all need forecasts of air transport activity.

The technical feasibility and the strategic master planning of an airport can often be examined adequately in the imprecise light of maximum likely forecasts. However, the more detailed design of facilities and financial feasibility require a set of forecasts which state the demand to an acceptable level of accuracy at a given point in time.

The accuracy inevitably deteriorates as the time-horizon extends, yet even 25 years may not cover the expected mid-life of the project if the planning process requires 7-10 years. It should be noted that the forecasts themselves are dependent on the level of technology, average aircraft size, income per capita, population, regional planning and a number of 'external' effects; in other words, the forecasts are usually implicitly 'locked-in' to the time frame for which they have been developed.

18.2 TYPES OF FORECAST NEEDED

The most fundamental forecasts are those for the gross annual uplift of passengers and freight for the system under consideration. This allows estimates to be made of the necessary scale of the system, its impact and, in gross terms, its financial viability.

For the more detailed planning and design purposes, the most crucial parameters are the flows of passengers, cargo and aircraft in the design hour, where the latter will normally be somewhat below peak hour flows, reflecting some acceptable level of delay or congestion in the system during the worst peaks. The passenger terminal layout and scale, and the revenue generation from concessions, are dependent not just on the total flow but also on the type of traffic, i.e. international/domestic, terminating/transit, charter/scheduled by class of cabin. These flows are usually derived from similarly classified annual flows by applying calibrated ratios which allow for the change in the peakiness with airport annual throughput. Typical ratios are given in Ashford and Wright (1992) and in Caves and Gosling (1999). An alternative approach is to construct likely future peak aircraft schedules and to derive the flows directly from these with assumptions on aircraft size and load factor. A similar method with different parameters is usually adopted for cargo terminals, a most important parameter here being the proportion of belly hold to all-freight cargo.

The total design hour flow consists of locally originating and terminating passenger and cargo flows, together with the passengers transferring (either on one airline or interlining between carriers) and the transit traffic which usually stays on an aircraft during a short turnround. Depending on the nature of the transfer traffic and the method of operating, it may be necessary to disaggregate it further for the design of customs and immigration facilities. Aggregation of all these traffic types allows an estimate of design hour air transport movements (atm) to be made, when combined with assumptions on the average design hour aircraft size and average design hour load factors. These aircraft movement data are essential for planning the airside capacity - runways, taxiways, aprons, gates. To complete the airside capacity planning, it is usually necessary to forecast General Aviation traffic, though its influence on the design hour capacity requirement is being reduced at large hubs, as it is progressively priced out of the market.

18.3 METHODS OF ANALYSIS

Standard texts on demand analysis quote many possible methods for assisting in the prediction of traffic (e.g. Ashford and Wright, 1992). The choice primarily revolves around the complexity and scope of the method. The more temporally stable and shorter term forecasts can usually be performed adequately with quite simple trend models which may not need to be unduly concerned with causality and uncertainty. This is not the case with long-term forecasts, which must concern themselves with unpredictable events both within and outside the air transport system. The choice is all too often determined by the availability of a data base and the budget for the study. Whichever method is chosen, the results must be understandable and presented in such a way that the users can exert their own informed judgement.

18.3.1 Informed Judgement

The simplest method of all, that of informed judgement, appeals because it needs very little data - indeed, too much data confuses the decision maker and slows the process down. However, the difficulties of making long term judgements are illustrated by a 1960s study by the Rand Corporation (Stratford, 1969), which predicted that by 2015 there would be only a 10 times increase in investment in computers and that it would be possible to control weather at a regional level. The recent Foresight study by the UK Office of Science and Technology (Loveridge, Georghiou and Nedeva, 1995) surveyed technical experts in their own fields. Among many quite reasonable predictions, it also concluded that multimedia teleconferencing would be preferred to business travel by 2007, that the direct operating costs of aircraft would be halved by 2008 and there would be autonomous aircraft that would not need air traffic control by 2007, these dates being the averages of the responses. Experts in the travel trade have also made some interesting predictions of events which influence transport demand and supply (Moutinho and Witt, 1995), including a 58 % probability of flying cars by 2020.

The Foresight exercise used a refined version of expert judgement called the Delphi technique, where a panel of experts have their judgements returned to them together with those of the other experts, so that they can adjust their views prior to the final collation of the results. The Delphi technique is also used by IATA in compiling airlines' views of the future.

Care should still be taken in using even the results of surveys of experts, since research shows that errors of individual judgements are systematic rather than random, manifesting bias rather than confusion. Further, many errors of judgement are shared by experts and laymen alike. Studies show that, at least in the field of fund management, the experts are no better than a random number generator at choosing investments. Erroneous intuitions resemble visual illusions in an important respect: the error remains compelling even when one is fully aware of its nature (Kahneman and Tversky, 1979). Most predictions, even when using analytic methods, contain an irreducible intuitive component, for example in the assessment of confidence intervals. Research shows that these are often assessed too optimistically, particularly before repetition provides the opportunity for feedback on predictive performance, because judgements are often made on minimal samples and assumptions are too restrictive of possibilities. A classic example of this is the tendency for long term forecasts to be unduly influenced by short term trends.

The more data are available, the more informed the judgement can be, at the expense in time and cost of incorporating them. It is sometimes possible to take data off the shelf, but there is usually something incompatible about them. They may well fall short in terms of sample size or the extent of their disaggregation. Also, they are very likely to be out of date. Thus, when methods requiring much data are chosen, a prolonged and expensive data collection exercise will generally be necessary.

Sometimes it is possible, rather than using ready-made data, to use ready-made forecasts, such as those produced by the aircraft manufacturers and governments (Airbus, 2000; Boeing, 2000; DETR, 1997), or analogy with a similar situation. Again, the problem is usually one of compatibility in time, space or characteristics of the setting. It may well not be appropriate to use available national forecasts to predict the traffic for a small regional airport. However, acceptable corrections can sometimes be made to suit the study in hand. Provided that adequate checks can be made on the validity of the base forecasts, this frequently can be the most cost-effective method.

18.3.2 Trend Extrapolation

It is usually necessary to generate some predictions which are unique to the situation under study. Further, the predictions have to be justified and therefore be based in some formal analysis of the historic development of the traffic. The simplest formal analytic technique is 'trend extrapolation', either in time or scale. Historic trends are derived by simple linear regression of the traffic itself or the annual growth rate. This is then projected into the future, modified by judgement to take account of changing circumstances. Short term trends often allow for the effect of economic cycles by applying some form of Box-Jenkins technique. Surprisingly, even the relatively simple data set required for trend analysis is not always available, or, if it is, it is not possible to discern a trend. Analogy or scenario writing then become invaluable techniques.

Longer term forecasts may recognise that the growth will mature over time by fitting the data to a Gompertz or 'S' curve. The difficulty over the long term is that small adjustments when fitting an S curve to historic data can result in widely differing estimates of saturation levels. Saturation can be estimated separately, to fix the size of the S curve, by, for example, assuming that the trips per capita,or 'propensity to fly' (ptf), eventually will not exceed those made in the US or Japan. This is a reasonable assumption since, regardless of how wealthy the travellers become, everybody has a limited time and energy budget. The relationship between air trips and wealth per capita is shown in Figure 18-1. It is, of course, necessary to compare like with like. The ptf on business for a small rural town is never likely to exceed perhaps 20 % of that for a capital city, though, for the same category of household, there might be nearly the same ptf for leisure. The ptf for the retired population may be very different from that for single parents.

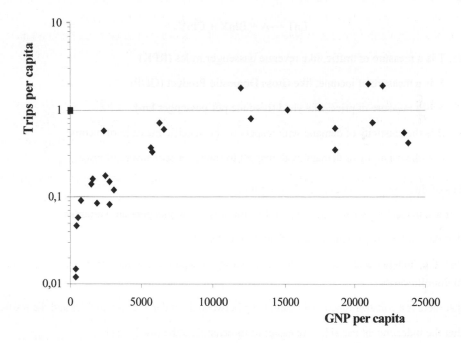

Figure 18-1   Propensity to fly

Another form of trend analysis is the 'step-down' procedure. This derives regional or local market shares from given national forecasts, making use of historic evidence of how the market shares have changed. The 'step-down' method is very popular in the USA, with its strong national forecasts and relatively stable traffic patterns, though it has been argued that the uncertainties created by deregulation in the distribution of traffic make it less appropriate now.

18.3.3 Econometric Models

If the above methods are deemed to be too naive, then it is necessary to bring causality into the analysis. The least complicated set of causal models is called econometric. These models try to relate the traffic to underlying economic parameters, or more readily available proxies for them. Usually they are calibrated by multiple regression of historic data to derive elasticities of demand, i.e. the change in demand due to a one per cent change in any one of the independent variables affecting the demand.

A typical econometric model might be:

$$LnT = -A + BlnX + ClnY$$

where: T is a measure of traffic, like revenue passenger miles (RPK)

X is a measure of income, like Gross Domestic Product (GDP)

Y is a measure of price, like yield (revenue per passenger km)

B is the elasticity of demand with respect to the specific measure of income

C is the elasticity of demand with respect to the specific measure of price

Models of this type carry a number of assumptions:

✈ that a satisfactory explanation is possible with only a few independent variables

✈ that the explanation is causal rather than co-incidental

✈ that the 'independent' variables are reasonably independent rather than suffering from multicollinearity

✈ that there is a constant functional relationship between the independent variables and the traffic

✈ that the independent variables are easier to forecast than the traffic itself

✈ that there are no significant errors in the data base.

Given that these assumptions can be accepted, the form of the model will depend on the particular circumstances. Multiplicative models are useful when the elasticities are essentially constant whereas exponential models are preferable when the elasticity varies with the independent variable. Difference models are to be preferred generally when the data set suffers from multicollinearity.

It is at least necessary to include terms to represent price and income, together with an autonomous trend term to describe changes over time in things like quality of service. A greater number of independent variables may increase the explanation of the traffic, but the greater the number, the more that have to be forecast. Also, it becomes increasingly difficult to interpret the model results.

The use of elasticities is shown in Table 18-1. The income variable used for leisure is disposable income and it is predicted to grow at between 2.0 % and 3.0 % per annum. Its associated elasticity is calibrated to be between 1.5 and 2.5. The price (yield) is presumed to fall by 1.0 % per annum and to have an associated elasticity of –1.0. There is also an autonomous trend term calibrated at between 0 and 1.5 % per annum. The consequent forecast is for leisure traffic to grow at between 4 % and 10 % per annum. Similarly, business traffic is expected to grow at between 3 % and 4 % per annum, the income indicator in this case being the value of trade and the traffic having a zero elasticity with respect to price. These elasticities are quite typical of developed air transport systems. Further examples of elasticities are given by ICAO (1986).

Table 18-1   The use of an econometric model

Determining Factors	Leisure	Business
Disposable income (% p.a.)	2 - 3	
Income Elasticity	1.5 - 2.5	
Trade (% p.a.)		4
Trade Elasticity		0.8 - 1
Real airline prices (% p.a.)	-1	-1
Price Elasticity	-1	0
Autonomous trend (% p.a.)	0 - 1.5	0
Potential growth (% p.a.)	4 - 10	3 - 4

Source: IATA Conference 1988, adapted by the authors

Econometric models may be applied to gross traffic, to route traffic or to generation of trips from a given geographic zone. Also they may be applied to classified sub-sets of the total potential travelling population in order to avoid the gross averaging process which is otherwise applied.

Any serious analysis of traffic at airports for marketing purposes, or to establish the viability of hubbing, really needs to work at the route-by-route level, and to understand the competitive situation in any origin/destination market. However, since the models are normally based in historic data, they deal only with demand which is revealed by the present behaviour patterns allowed by the historic characteristics of the transport offered. Also, there are many other aspects of the travellers' behaviour which contribute to their travel decisions than simply the economic characteristics of the subject modes.

18.3.4 The Travel Decisions

To begin to model behaviour in more detail, it is necessary to consider the need to travel, the mode of travel and, if the mode is air, which airport the traveller will use. The generation element of the decisions may be dealt with by market category analysis techniques as used by Roskill (1971) or by econometric modelling. The trip generation per capita varies a great deal between industrial sectors and also by socio-economic group, with employees of small high technology research and development companies in the US making 150 trips per year while the average for UK citizens is one per year. In conventional transport planning, estimates of trip generation would be followed by distribution models, usually based on some analogy to the laws of gravity. This method presupposes that there is a reservoir of generated demand waiting to travel to any available destination. Simple models of this type ignore the important contribution of specific 'community-of-interest' between the origin and destination zones. A technique which distributes traffic as a function of relative attractiveness of the destination helps to overcome this drawback. Generation and distribution models can be applied to the air mode in isolation but, for shorter haul situations, it is more appropriate to model total generation and distribution and then to derive the modal split to air by use of a modal choice model.

18.3.5 Modal Shares

The availability of high speed rail services will certainly reduce air traffic on dense short haul routes, as demonstrated by the TGV experience. Airbus Industrie has suggested that a one hour advantage increases market share by 20 per cent (Reed, 1990). The Japanese experience, with quite similar fares for air and rail, is that air begins to take a greater share of traffic than rail at about 750 km, as shown in Table 18-2. Certainly there is a switch of even high yield traffic from air when door-to-door time by rail becomes less than three hours, but interlining and natural growth in traffic can still leave a significant air market: Orly to Nice, Orly to Marseille and Orly to Brussels remained as three of the densest routes in Europe (de Wit, 1995) despite the high speed rail opportunities, and BA's shuttle from Heathrow to Manchester has more than a million annual passengers despite trains averaging 160 km per hour over only 300 km. The resilience of air is due partly to the diffused nature of the demand within metropolitan areas (at least 50 per cent is home-based rather than office-based) and partly to the access to down-town rail stations being often more difficult than to airports. Less than 10 per cent of the European scheduled airline capacity is threatened by a future high speed rail network (Veldhuis et al, 1995). None-the-less, there are routes where rail does compete well, and in these cases it can certainly serve to reduce the strain on air capacity.

Table 18-2   Air and rail shares in Japan

A: **Shares of total trips**

Mode	300-500 km	500-750 km	750-1,000 km	1,000 km +
Air	4.5 %	13.9 %	50.0 %	83.6 %
Rail	48.8 %	64.8 %	37.3 %	13.2 %

B: **Comparison of fares and journey times**

Route	Rail - air fare (Yen)	Rail - air time (hr:min)
Tokyo - Osaka	3 930	1:30
Tokyo - Fukuoka	3 780	3:23
Tokyo - Yamagata	1 990	1:32
Osaka - Sendei	4 175	3:04
Nagoya - Sendei	4 860	2:15
Fukuoka - Niigata	5 975	5:09

Source: Air Transport World, July 1995, pp. 24-32, Adapted by the authors

Telecommunications may be able to replace some physical travel, but it is generally accepted that, over short haul distances, any substitution effects are compensated by stimulation of travel (Salomon, 1986). Videoconferencing on the desktop is likely to become very common. A major Swiss bank has installed 75 units worldwide, their travel department applying a 'could it be done by videoconference' test (Arvai, 1993). The greatest impact in the US is expected to be in intra-company connections, where some 30 per cent of air trips may be substituted by 2010 *(Aviation Week, 29 November, 1993, p 40)*. The future threat from telecommunications is more likely to affect long haul air travel. The differential advantages of teleconferencing must be greater at long distances, particularly with falling prices as the large capacity of fibre optic cables comes on stream. Telecommunication cost is almost constant with distance. The danger for aviation would be that a relatively small impact on the high yield business market might exert a pressure for increased fares in the highly elastic leisure market.

18.3.6 Discrete Choice Models

Attempts to increase the richness of the behavioural content of models are often made by using a disaggregate, or discrete, approach; involving the knowledge and modelling of individuals' trips. This type of modelling is usually more accurate in determining responses to changing supply or choices between alternatives, because it retains the maximum richness of the available information, rather than hiding some of the variety through the aggregation process. It is more important to use disaggregate information to calibrate a model: The application can often be satisfactory on aggregate data.

Most of the attempts to model the revealed choice behaviour of passengers have used individual trip data to calibrate logit models, because of the models' specific properties (Fischer, et al, 1990; Hensher and Johnson 1981). They express the probability of choosing an alternative in terms of the ratio of the utility to the passenger of any one of the alternative choices to the combined utility of all available alternatives, where the utility function to be maximised is a function of the attributes of the alternative and of the traveller. Almost always, the best explanation of behaviour in choosing between airports, between routes or between carriers has been obtained by using logit models with some combination of access time, trip cost and frequency as the dependent variables.

Table 18-3   Choice of London airport, percentage by reason

Reason for choice of airport	Gatwick	Heathrow	City	All airports
Flights available	45.1	24.1	9.4	31.0
Nearest to home	13.4	11.4	20.0	13.3
Decision made by someone else	8.0	14.0	5.2	11.5
Connecting flights	3.3	11.7	0.1	8.3
Prefer airline	6.4	9.6	0.1	8.0
More economic / cheaper	10.4	3.5	1.3	6.0
Timing of flights	1.9	7.4	7.9	5.4
Nearest to business location	1.3	5.8	26.8	4.3
Total passengers (000s)	17 990	36 887	171	58 357

Source: UK Civil Aviation Authority, CAP 610, 1991, adapted by the authors

The most important choice factor, after the availability of the service, is the ease of access to the service, as shown in Table 18-3. Other factors influencing the choice decisions are the nationality mix of the airlines, the aircraft technology and, of course, fares, though it is difficult to obtain data on the fare actually paid. The effect of replacing turboprops with jets on competing routes from airports in the Midland region of the UK was estimated by comparing turboprop predictions with jet outcomes (Brooke et al, 1994). The improvement in market share due to jet service appeared to be less than 10 %; indeed, it may have been less, because no attempt was made to disassociate the changes in type from the increase in aircraft size of the jets (85 seats, rather than 50-65 seats) and many in the airline industry believe that it is capacity rather than frequency which defines market share. However, other evidence suggests that a balanced market with both operators using turboprops would change to 80 %/20 % if one operator used jets at the same frequency (Aroesty et al, 1990), and this seems to be confirmed in the experience of introducing the new 50 seat regional jets.

18.3.7 Revealed and Stated Preferences

The data for disaggregate models is usually obtained from surveys of historic travellers; i.e. they are based on revealed demand. This clearly cannot account for those who have not travelled, nor for those whose travel decisions have been significantly constrained by the choices available to them: for example, charter passengers' departure airport is often constrained by flight availability. At the risk of irresponsible answers, these difficulties can be overcome by the use of 'stated preference' surveys. There are techniques for maximising the relevance of the answers obtained, the best of which are probably to impose budget constraints and to check against actual behaviour. The prime advantages of the technique are the cheap and rich nature of the data and the ability to consider larger scale variation in the alternatives than have actually been tried in real situations.

18.3.8 Effects of Supply Decisions

The demand can only be expressed if a service is offered. It is becoming increasingly common for supply decisions to be uncoupled from revealed demand. This is sometimes to correct an under supplied market, but it is often a result of airline or airport strategies. A recent example is the KLM initiative to markedly increase frequencies and capacity at Amsterdam. The traffic may then appear at points in the system that are quite different from those of interest to the passenger or shipper. It must be anticipated that many other, and more extreme initiatives will be taken as the possibilities of liberalisation of services in Europe are realised. Also, liberalisation is continuing to spread through the rest of the world.

18.3.9 Uncertainty

The calibration of those models which attempt to introduce causality into the understanding of passenger behaviour require a great deal of data on the mode characteristics and the demography of the potential travellers, as well as richer socio-economic information than is generally used in simpler econometric modelling. All the models' predictive ability also relies on being able to forecast the future trends of the variables used. Because of these difficulties, prediction based in any of the above methods suffers from likely errors which become greater as the time horizon increases. It has been said that a central case forecast is simply that outcome with a 50 per cent likelihood of being wrong in either direction.

Figure 18-2   Market research requires various techniques of data collection, photo A.Kazda

There seem to be four ways in which this problem of uncertainty can be approached. The most common approach is to use sensitivity analysis on all the factors which it is felt may be in error. A rather more sophisticated approach is to assign probabilities to the forecast of the independent variables and let the output take the form of a risk analysis. In this case no preferred forecast is offered, rather the forecast's uncertainties are made transparent to the decision maker. It should, however, be noted that there is much implicit judgement in the assignment of the probabilities, just as there is in all the quantitative methods. It is not only the purely judgemental methods that require judgement from the analyst. A third approach is the normative one of adopting policies to limit or encourage demand - in this case the models' role is to provide the understanding necessary for setting the levels of policy variables to adjust demand to achieve the planning objectives. The fourth approach is to recognise that it is not possible to predict the future, but that it can be explored by writing scenarios and interpreting the consequences.

18.3.10 Scenario writing

Scenario writing allows potential futures to be described. The forecasting techniques can then be applied to explore the traffic implications of each of those scenarios. Unfortunately, scenario writing is often used suboptimally, being limited to exploring options within the system under consideration with a view to predicting the future probabilistically. However, a more productive use is to explore the range of feasible potential futures and their consequences in terms of the needs to which a system might be asked to respond and the steps it would need to take in order to respond effectively. It should be emphasised that the objective is not to guess the future, even by assessing probabilities of the various potential futures and hence take a view on the most likely future. Scenarios explore potential futures so that some light can be thrown on the scope and flexibility which needs to be designed into the system, and on the consequences for the system's performance of not being able to meet the needs of some scenarios. Alternatively, the process can identify those possible futures which the system should not be designed to accommodate.

The essence of the method is to identify those factors which drive the business, but over which the business has little control, and condense them into several major themes. Once the consequences of the themes have been quantified using the conventional forecasting techniques, strategies may then be devised to meet each of the separate scenarios. Finally, the strategies are merged into a core strategy which is capable of responding to any one of the scenarios.

It can be taken for granted that the most likely future is the one which is presently being projected by in-house planners and system designers. It can also be taken for granted that, except in so far as the system's future use is predetermined by the closing off of future options by sunk investment, the most expected future will not coincide with any actual future state. None-the-less, this 'business as usual' case is an essential scenario, because it forms the base from which the relative implications of the other optional scenarios can be explored.

There are many options for writing scenarios. The important characteristics are that the scenarios should be cohesive and should show a feasible route from the present state to the potential future state. The latter requirement is made difficult because the dynamics of change are not well understood (Keynes quoted in Wills, 1972, p.165).

One example of the approach is to paint four potential scenarios (high and low growth, egalitarian or inegalitarian) for each of three (conservative, reformist, radical) ideologies, and to ponder the implications for international trade, political and economic stability and the need for transport. It can, for example, be inferred that a conservative high growth egalitarian society will generate technological sophistication in the northern hemisphere, and the encouragement of private transport, including aircraft. A reformist low growth more equal society would emphasise conservation and would encourage innovation in operating systems rather than in the technology of transport. The Dutch have used a consolidated version of these scenarios to inform their planning for Schiphol airport (Ashley et al, 1994).

Once inferences have been made within each of the scenarios of the likely associated impacts on the factors driving the econometric models, they can be used to quantify the implications of the various potential scenarios, in the same way that the models would normally be used for the prediction of expected traffic. It may be that one of the scenarios would lead to a halving of the cost per pkm, in which case the consequence would be a doubling of demand if the price elasticity consistent with that scenario was judged to be 2.0. Note that the traffic itself may or may not double, depending on other consistent inferences in the same scenario with respect to available capacity, income, etc.

18.4 HISTORIC TRENDS IN TRAFFIC

It is clear from the above discussion that much conventional forecasting activity relies on a knowledge of past trends: with some justification, because only an unfortunate coincidence would place the forecaster at the exact moment in time where a complete discontinuity occurred.

The gross uplift has grown approximately logarithmically over the long term. Recently, there has been a slowing of growth in the more mature markets with the inference of a swing towards the third world in future demand, a rise in importance of freight, a fall in mail and a slower domestic growth. The growth in air transport movements historically has been less than that in passenger or tonne kilometres because of a trend towards longer stage length, larger aircraft and, latterly, a higher average load factor. Recently, this has been countered by an emphasis on frequency in liberalised markets, by over-water twin-engine aircraft and regional jets of 50 seats and less.

Non-scheduled traffic, at least in Europe, has been much more sensitive to changing economic circumstances than scheduled traffic.

18.5 FACTORS AFFECTING THE TRENDS

18.5.1 Economic Factors

In the more complex and data-hungry models, gross or discretionary income per capita tends to be used as the main causal variable. In more macro models, Gross National Product (GNP) is normally used as a surrogate for income. Modellers go to considerable lengths to interpret government views and to form their own views of future world events, but there tend to be dominated by conventional wisdom operating on common data bases. One must ask whether GNP is really any easier to forecast than traffic growth, particularly in the short term. Also, one must ask whether the influence of GNP is all-pervading as indicated by simple correlation analyses at the macro level. The growth of traffic over time in Brazil suggests a more complicated relationship, with a relationship between the change in the propensity to fly and the change in GNP per capita which passed through levels of 5.5:1, zero and 3.5:1 as the economy has moved from pre-industrial through the formation of an industrial base and finally to development. It is often difficult to get sufficiently local income data to be of much use for smaller airport studies.

Other important economic factors include exchange rate differentials which drive all types of travel, as exemplified by the changes in North Atlantic traffic as the dollar weakened in the late 1980s (Avmark Aviation Economist, March 1994, p 11). These are notoriously difficult to forecast but make a major difference in the competition between holiday destinations, as does the threat of terrorism. There are also likely to be shifts in the global distribution of economic activity and in the distribution of wealth both globally and within nations.

18.5.2 Demographic Factors

Conventionally, air traffic increases with population, urbanisation, reduction in size of households and with younger age groups. It is generally assumed that these variables are amongst the easiest to forecast and that their relationships with traffic growth are stable. However, forecasts of the USA population in 2010 have varied since 1945 from 291×106 to 381×106 as the births per 1000 women have changed from 85.9 (1945) through 122.9 (1957) to 73.4 (1972). Again, the tendency for an increased propensity to fly (PTF) may currently be explained largely by income, household make-up and job-type factors. It is quite possible that these relationships may break down. The changing structure of industry and commerce could affect air trip generation rates. Propensity to fly (ptf) varies greatly with the type and size of firms. There is evidence of a reversal of domination by the top 500 Fortune companies in the US (Makridakis, 1995), which could have a strong bearing on the future ptfs.

There is already a tendency to decentralisation of metropolitan areas (New Scientist, 15 June 1996, p 11), and this may lead to greater demand for lower density air travel.

Some feel that the information revolution, or 'fourth logistic revolution' (Anderson and Batten, 1989), will result in fast and volatile flows of commodities, people and information, leading to a preference for the fastest and most direct mode of travel. This type of future, where a new hierarchy of cities evolves with the most powerful being those based in the four Cs of Culture, Competence, Communication and Creativity, is already visible. It should be put alongside other demographic trends, which include the labour migration flows which eventually form the basis for Visiting Friends and Relatives traffic (International Labour Rights, Second Quarter, 1993), and changes in fertility and mortality.

18.5.3 Supply Factors

Cost per passenger kilometre (pkm)

At constant levels of regularity, reliability, frequency and comfort, the primary supply factors determining demand are ticket price and generalised cost (i.e. a combination of price and value of time saved). The price is determined by the underlying cost of production and the pricing policies adopted to recover the costs. The costs of production are, in turn, determined by operating policies with respect to load factor, aircraft size, input factor costs, productivity and advances in technology. The load factor in turn depends on price so that the revenue yield per seat falls as load factor rises. Estimates of world-wide load factors reveal an increase at about 1/2 per cent per year from 55 per cent to 65 per cent by 1995. Aircraft size had been expected to increase on the densest routes as most of the forecast demand is taken up by size rather than frequency in order to reduce costs in the competition for the low fare passenger and to combat the effects of congestion. In these cases, annual growth in seating capacity would have been almost the same as passenger growth. In the event, even on these dense routes, frequency competition appears to be overcoming any shortage of runway capacity. On lower density routes, and where increasing competition is unhindered by congestion, increased frequency may well cause even higher ATM growth rates. Meanwhile, ever more powerful revenue management systems have allowed airlines to sell increasingly marginal seats, so increasing load factors with little increase in revenue per flight. Some airlines, notably BA, are reducing capacity in order to reverse this trend of excessive yield dilution.

Cost of input factors

Fuel has dominated the historic changes in input factor prices. A combination of shortages and the long lead times for aircraft technology to respond to the situation has, from time to time, caused considerable problems for the industry. Operational changes have helped to alleviate the situation, but in the longer term the larger cost reductions to combat forecasts of price increases of 1 per cent per year will have to come from the technology. There is little sign of an economic incentive to invest for this level of improvement. The combined effect improvements in of labour costs, advances in technology and reductions in the first cost per seat of aircraft are unlikely to result in a total cost reduction of 1 per cent per annum.

Fares

Data on real air fares paid are usually only available on an aggregate basis for all operations of an airline, quoted as yield per pkm. Forecasts of changes in yield may then be used to estimate traffic by using fare elasticities. The derived historic elasticities may be modified as necessary to allow for changes in local conditions. It should be noted that none of these supply costs, fares or elasticities are necessarily appropriate for other situations than those for which they were derived - for example, Jung and Fujii indicate that short haul price elasticity in the USA may be as high as 2.74. Some estimates of future fares are given in Table 18-4.

Table 18-4   Air fare assumptions for two growth scenarios (%change per year)

Demand	Short haul				Long haul				Domestic	
	Leisure		Business		Leisure		Business			
Scenario	Low	High	Low	High	Low	High	Low	High	Low	High
1995 - 2000	2.9	-0.2	3.7	0.6	1.4	-0.7	0.6	-1.5	4.0	2.4
2000 - 2005	2.0	0.1	2.1	0.4	0.6	-1.2	0.0	-1.7	1.8	0.5
2005 - 2015	1.7	-1.2	0.3	-0.9	0.2	-1.2	-0.3	-1.7	0.0	-0.3

Source: UK Department of Environment, Transport and the Regions, web site

18.5.4 Technology

The impact of technology on future capacity is often the most neglected area of supply prediction. Heathrow's growth is a case in point, early predictions being hampered by inhibitions on the maximum feasible size of aircraft and more recent analyses by under prediction of annual runway capacity. Despite the highlighting of these broader system implications in the literature (e.g. TRB,

1990), most studies only allow for changes in a single factor (.e.g regional jets) which is closest to the interests of the immediate focus. In fact, it is clear that there will be changes in navigation, communication, guidance on approach and on the ground which will have implications for capacity, as well as changes in vehicle technology which will affect both cost and capacity. It is less clear if and when specific changes will be introduced, though they will only occur when there is a sufficient incentive.

18.5.5 Management

The technical changes will often open up new opportunities for changes in management style, as with new routes and networks for airlines, new operating procedures around airports, advanced revenue management systems and improved facilitation through the use of smart cards. However, managers can also innovate without technology, as is seen by the growth of the low cost carriers and the forming of global alliances. These and other initiatives will continue, but which of them, when and in which combinations can only be explored by scenario writing.

18.5.6 Regulation

Since the deregulation of US domestic air transport in 1978, the ethos has spread through Europe and many other countries. It has been accompanied by a parallel trend to the privatisation of airlines and, later, of airports. Without these trends, there is less opportunity for competition, for fare reductions, for route development, for alliances, for efficiency gains and for airports to take marketing initiatives. There have been abuses of the monopoly positions that have been made possible by the new freedoms. Rather belatedly, this has led to the imposition of regulatory controls on monopoly powers. The extent to which this 're-regulation' will succeed in releasing more effective competition, and to which the experience will influence the policies of countries that have not yet liberalised and privatised, is difficult to predict.

The regulatory setting should therefore be incorporated in scenario descriptions, including free trade, open skies, foreign ownership and environmental taxes or capping as well as the regulations to control monopolies.

18.5.7 Constraints

The above factors will only influence future traffic in the expected way if sufficient investment is made to accommodate the potential demand. Therefore, any political decisions could interfere with forecasts, as could any ineptitude in the planning process. In an ideal forecasting process, these effects would be internalised within the method, but political forecasting is not yet sufficiently advanced to allow this.

18.5.8 Cargo

The analysis of the cargo market is similar in basic method and generation variables. Some additional factors affecting cargo growth are:

✈ continuing increased share of integrated carriers

✈ time and cost sensitive products

✈ early stage of a product's cycle

✈ availability of capacity on routes

✈ regulation of facilitation

✈ competitive advantage over other modes

18.6 CONCLUSIONS

The discussion has identified the requirement for forecasts of arriving and departing passengers disaggregated by purpose and type of destination, and of air transport movements by type of equipment. These forecasts are required on an annual basis but, more importantly, for the representative peak hours. Information is also needed on General Aviation and military activity, but the impact of this traffic on design hour movements is more a function of airport policy than of the unconstrained volume of traffic.

The methods available to make these forecasts have been reviewed and shown to be more or less adequate for short term detailed planning and design, but to offer little more than a formal basis for reasoned judgement in the long time horizons needed for airport system and master planning. Interactions between the variables which control demand are not sufficiently understood to allow more certain forecasts to be developed, so that informed judgement must be used. This may be expressed explicitly as high and low forecasts, or it may be combined with formal attempts at risk analysis, or it may inform the interpretation of future scenarios.

It is clear that there should be consistency between airport forecasts. The obvious way to obtain this consistency would be by modelling the complete air transport net with a single methodology, a single set of assumptions and a consistent set of data. However, not only is it rare to find the resources for this, but there is a real danger of constraining the output to that which follows from a dominant preference of one member of the team of analysts and advisors. There is no substitute for keeping an open mind on the future, and exploring a wide range of possibilities. Consistency of methodology is undoubtedly necessary for communication and interpretation, but the inputs to the prediction process should be allowed to vary as necessary to reflect the variety in the system.

No particular model will always be the most appropriate, nor will any one model always give the most accurate prediction. There is no substitute for using more than one approach. Much can be learnt from a comparison of the outcomes relative to the constraints of the modelling. However, the balance of cost and competence will often be best achieved by a simple econometric model, applied to a carefully selected set of markets that most affect the total traffic. If necessary, these models can be used even without a good historic data set for calibration, since, with care, appropriate elasticities can be borrowed from other studies.

Many of the important imponderables are external to the air transport system, and are best dealt with by incorporation in a range of comprehensive scenarios. Inside the system, many other factors over which most airports have little control have been identified in Section 4, and these also should be included in the scenario descriptions.

Apart from economic growth, the factor which strongly influences the rate of growth of demand in the short term is the airlines' and the regulators' policies on fares. The conventional wisdom sees a continuation of a low fares policy fuelled by deregulation, leading to high load factors and larger aircraft. If this approach leads to insufficient yield to at least maintain the present airline rates of return on investment and hence leads to lack of capital to fund the needed extra capacity, the airlines may retrench and rely on the essential business traveller. If this happened, too much surplus capacity would have been provided. The low cost carriers can make good use of some of this, but the US experience is that only the most resilient survive.

The development of technology, both to combat fuel availability and price and also to counter the physical and environmental constraints on airport development, is important in the longer term. Further interruptions in the oil supply may increase the rate at which more fuel-efficient aircraft are developed, and flight frequencies are reduced so increasing the passengers per aircraft. Advances in aircraft technology and approach guidance, designed to allow changes in operational practices, will be important in lifting the capacity constraints at large airports. Their implementation depends partly on whether the large global carriers really want capacity to be increased at their hubs.

One other air transport factor which is particularly important in determining an individual airport's traffic is the extent that the network develops as a hub-and-spoke system rather than concentrating on direct connections, and which are chosen as hubs or as spokes or are left out completely. To some extent, the airports will influence this, since the hub system will tend not to develop if transfers are not convenient.

Despite all these caveats, forecasts are still needed, so one current view of the future is now given in Table 18-5. A low fare policy may result, in the short term, in emplanement growth rates higher than those forecast as mean values for European airports. Other expected trends are for ATMs to grow more slowly at larger airports and for cargo to grow faster than passengers. The forecasts will depend on conditions at individual airports in terms of level of maturity, surface competition, marketing and the airport's share of the national traffic.

Table 18-5   Annual average predicted growth in passenger kilometres, 1999-2018

Traffic Growth to/from:	Africa	Asia	Europe	Middle East	Latin America	North America
Africa	4.9	5.4	4.9	4.0	6.9	6.7
Asia-Pacific	5.4	6.4	6.1	4.5	4.2	5.6
Europe	4.9	6.1	4.3	3.9	5.5	3.7
Middle East	4.0	4.5	3.9	4.6		3.9
Latin America	6.9	4.2	5.5		6.9	4.7
North America	6.7	5.6	3.7	3.9	4.7	2.9

Source: Boeing Aeroplane Company web site, adapted by the authors

In lieu of any specific study for a particular airport, the gross forecasts may be used to factor up from base data, with the caveats mentioned in this section. The peak hour ATMs may be estimated by making assumptions about the peak hour fleet mix and load factor. These last two factors will always depend very much on the circumstances at any particular airport, not least the amount of transit traffic on the aircraft.

FURTHER READING

ACI (1996). *Apron Safety Handbook.* Airports Council International, Geneva.

ACI Europe(1995). *Environmental Handbook.* Airports Council International, European Region, Brussels. January.

ACI Europe(1993). Technical & Safety Committee: *Winter Service White Booklet.* ACI Bruxeles.

ACI Europe (1999). *Winter Services Yearbook.* Airports Council International-Europe, Brussels.

ACI ROUNDUP (1992, 1994). N°2: *Letting the train take the strain.* Airports International. January/February, March.

Airbus Industrie (2000). *Market Perspectives for Civil Jet Aircraft.* http://www.airbus.com/gmf99_1.html.

Airport Support. (1991). *Cashing in on Concession.* Trading Supplement. April.

Airport Support Trading Supplement. (1991) *The Cost of Banning Duty Free*, April.

Anthony, R.R., Lindsay, C.H., Oke, J.L., Stevens, A:B. (1981). *Fire Extinguishants: Their history, properties and use.* ICAO Bulletin. October.

Arvai, E, S. (1993). *Telecommunications and business travel: the revolution has begun.* 8th International Workshop on Future Aviation Activities, pp28-31, Transportation Research Board, National Academy of Sciences, Sept. 13-15.

Ashford, N., Bolland, S., Ndoh, N.N. (1993). *Planning Surveys for Modeling Airport Access*, Air Transport in Central Europe Conference, Jasná – Slovakia.

Ashford, N. (1993*). Current Development in European Airports.* Loughborough University of Technology, Loughborough.

Ashford, N., Moore, C.A. (1992). *Airport Finance*, Van Nostrand Reinhold.

Ashford, N., Wright, P.H. (1992). *Airport Engineering*, 3 rd, Wiley Interscience, New York.

Ashford, N., H. P.M. Stanton and C. A. Moore. (1995). *Airport Operations*, McGraw-Hill.

Baier, V.M. (1993). *Flughafen Frankfurt/Main AG Well Prepared to Weather at Severe Winter.* ACI Winter Service Workshop, Budapest.

Bei, B. (1992). *The Growing Importance of Trading to the Quality of Airport Service*, First Airport Trading Conference AACI Europe, Paris.

Birlie, K. (1993). *Aircraft de-icing*, 2nd ACI Europe Winter Service Workshop, Budapest.

Blow, C. (1996). *Airport terminals.* 2nd ed., Architectural Press, Oxford.

Boateng, E.A., Schweizer, F. (1994). *Disposal for Airport Waste, New World Transport '94.* Sterling Publishing Group PLC, London.

Boeing Commercial Airplane Group (2000). *Current Market Outlook.* http://www/boeing.com/commercial/cmo/5apc1.html.

Brooke, A. S., R. E. Caves and D. E. Pitfield. (1994*). Methodology for predicting European short-haul air transport demand from regional airports.* Journal of Air Transport Management, Vol. 1, No. 1, pp 37-46.

Buffet, A. (1987). *Jet Fuel Distribution.* ITA Magazine N°42, Paris.

Butterworth-Hayes, P. (1991). *Learning the Lessons of a Real Disaster.* Jane´s Airport Review. January/February.

CAA. (1989). *CAP 556 – Passenger at the London area airports in 1984.* London.

CAA. (1981). *CAP 384 – Bird Control on Aerodromes.* London.

Caves, R E; A D Kershaw and D P Rhodes. (1999). *Operations for airport approach noise control - flight procedures, aircraft certification and airport restrictions*, Transportation Research Record 1662, pages 48-54.

Caves R. E. and N. N. Ndoh. (1995*). Investigating the impact of air service supply on local demand - a causal analysis.* (with Environment and Planning) A, Vol. 27, pp 489-503.

Caves, R. E., L. R. Jenkinson and A. S. Brooke. (1996). *Determinants of emission from ground operations of aircraft.* Paper presented to 20th Congress, International Council of the Aeronautical Sciences, Sorrento, Italy. September.

Caves, R. E. and G. D. Gosling. (1999). *Strategic Airport Planning*, Elsevier.

Collins, A. and A. Evans. (1994). *Aircraft noise and residential property values.* Journal of Transport Economics and Policy, May, 175-196.

Coogan, M. (1995). *Comparing airport ground access: a transatlantic look at an intermodal issue.* TR News 174, November-December, 2-10.

Československé aerolínie. (1983). *Technické podmínky pro používání močoviny (Karbamidu) na letištích ČSA, (Technical conditions for urea using)* PZ-A1-02/36-LT, ČSA Praha.

Čihař, J. (1973). *Letiště a jejich zařízení I. (Airports and its facilities I)*, Alfa Bratislava,

Bratislava.

Čihař, J. (1975). *Letiště a jejich zařízení II. (Airports and its facilities II)*, Alfa Bratislava, Bratislava.

ČSN 650201: *Horľavé kvapaliny – prevádzky a sklady (Standards for flammable liquids)*

ČSN 757221: *Klasifikace jakosti povrchových vod (Standards for ground water quality)*

ČSN 830915: *Ochrana vody pred ropnými látkami – objekty pre manipuláciu s ropnými látkami a ich skladovanie. (Standards for ground water protection against petroleum products).*

D'Albiac (1957). *London Airport*. Journal of the Royal Aeronautical Society, 61, April, 225-237.

Deasy, C M. (1974). *Design for human affairs*, Wiley, New York.

Dempster, D. (1989). *Standardization Needed for Visual Docking Guidance*, Airport Forum. January.

De Neufville, R. (1976). *Airport System Planning,* The Macmillan press Ltd, London.

DETR (1997). *Third party risk near airports and public safety zone policy.* (A report to the Department by consultants) UK Department of the Environment, Transport and the Regions, London, October.

DETR (1997). *Air Traffic Forecasts for the United Kingdom*, UK Department of Transport, Environment and the Regions.

DETR (1999). *Guidance on Airport Transport Forums and Airport Surface Access Strategies.* UK Department of the Environment, Transport and the Regions.

Doganis, R. (1991). *Flying off Course – The Economics of International Airlines*, Harper Collins Academic.

Doganis, R. (1992). *The Airport Business*, Routledge London.

Drahota, J. (1990). *Letecký hluk, (Aircraft noise).* Letecký obzor, June.

Dudáček, L. (1988). *Záchranná a požární služba. (Emergency services),* Letecký obzor. February.

Duff, A. (1999). *Access – the Heathrow Case Study.* Loughborough University.

Easthill, A. (1987). *Mobile Fuel or Hydrants*, Airports International, February.

EC. (1996). *Impact of the Third Package of air transport liberalisation measures.* European Commission, Brussels.

FAA. (1992). *FAA Integrated Noise Model Version 3.1.* DOT/FAA/EE-92/02 Office of Environment and Energy Washington DC, June.

FAA. (1996). *FAA Integrated Noise Model Version 5.1.* FAA-AEE-96-02. Office of Environment and Energy Washington DC.

Fábera, J., and M. Kyncl. (1977). *Dopravní letiště (Transport airports)*, NADAS. Praha.

Fergusson, M. (1995). *Environment and Air Transport in Europe.* In conference: Environmental Aspects of Air Transport, Royal Aeronautical Society, London, 19 September.

Fischer, M.M., P. Nijkamp and Y.Y. Papageorgiou, eds (1990). *Spatial Choices and Processes*, Elsevier Science Publishers B.V.

Gabriel, P. (1985). *Ornitologická ochrana letiště Kunovice, (Ornithological protection of Kunovice airport).* VŠDS Žilina, Žilina.

Gangl, S. (1992). *The Financial Importance of Airport Trading in the Creation of Airport Capacity.* First Airport Trading Conference AACI Europe, Paris.

Gibson, B. (1992). *The Role of High Street Names in Airport Trading,* First Airport Trading Conference AACI Europe, Paris.

Gosling, G. D. (1997). *Airport ground Access and Intermodal Interface.* Transportation Research Record 1600, pp. 10-17.

Gould, R. (1994). *Runway De-icer Revolution Gathers Pace,* Jane's Airport Review, April.

Graham, A.. (2000). *Demand for leisure travel and limits to growth.* Journal of Air Transport Management, Vol. 6, No. 2, pp. 109-118.

Graham, A. (1991). *Airport Economics.* Polytechnic of Central London, London.

Gray, F.J. (1991). *Developing non-aeronautical Income*, Sypher: Mueller International Inc., Loughborough University of Technology, Loughborough.

Grover, J.H.H. (1990). *Airline Route Planning*, BSP Professional Books.

Halse, J. (1999). *Concessions space, lecture to Short Course in Airport Airside and Landside Design*, Dept of Civil and Building Engineering, Loughborough University, November.

Hill, R.E. (1990). *Airport Security Beyond the 1990s,* Closing the Gate on the Terrorists, Airports into the 21st Century Conference, Hong Kong.

Hill, R.G., Howell, W.D., Sarkos, C.P. (1982). *Full-scale Wide-body Test Article Employed to Study Post Crash Fuel Fires*, ICAO Bulletin, October.

Honeybun, M. (2000). *Fuel Supply to Airports,* International Airport Review, Vol. 4, Issue 1, pp. 17-24.

Hopkins, H. (1985). *Cold Cures in Helsinki,* Flight International, 12 October.

Horonjeff, R. and F. X. Mckelvey. (1994). *Planning and Design of Airports*, 4th ed. McGraw-Hill, New York.

Hudson, G. M. (1982). *Airport Environmental Planning*, Loughborough University.

Hughes, C.V. (1990). *More Foam Needed*, Jane's Airport review.

IATA/*: (1992). *Air Transport & The Environment*, IATA, Geneva, /* International Air Transport Association.

ICAO (1984). *Aerodrome Design Manual*, Part 1, Runways, Doc 9157-AN/901, 2nd ed.

ICAO Journal (1990). *Air transport in the European Community. The hardcore problem.* December.

ICAO (1997). *Outlook for air transport to the year 2005,* Circular 270-AT/111, International Civil Aviation Organisation, Montreal.

ICAO (1986). *Manual on air traffic forecasting,* 2nd ed. Doc. 8991 - AT/722/2, International Civil Aviation Organisation, Montreal.

ICAO (1999). *Aerodromes – Annex 14 Volume I Aerodrome Design and Operations*, Montreal, July.

ICAO (1990). *Airport Services Manual, Part 1 – Rescue and Fire Fighting*, ICAO Doc 9137-An/898, Part 1, 3-rd edition, Montreal.

ICAO (1984). *Airport Services Manual, Part 2 – Pavement Surface Conditions,* ICAO Doc 9137-An/898, Part 2, 2nd edition, Montreal.

ICAO (1991). *Airport Services Manual, Part 3 – Bird Control and Reduction,* ICAO Doc 9137-An/898, 3-rd edition, Montreal.

ICAO (1986). *Airport Services Manual, Part 5 - Removal of Disabled Aircraft*, ICAO Doc 9137-An/898, part 5, 2nd edition, Montreal.

ICAO (1983). *Airport Services Manual, Part 6 – Control of Obstacles,* ICAO Doc 9137-An/898, 2nd edition, Montreal.

ICAO (1980). *Airport Services Manual Part 7 – Airport Emergency Planning,* ICAO Doc 9137-An/898, 1st edition, Montreal.

ICAO (1984). *Airport Services Manual, Part 8 – Airport Operational Services*, ICAO Doc 9137-An/898, Part 8, 1st edition, Montreal.

ICAO (1984). *Airport Services Manual, Part 9 – Airport Maintenance Practices*, ICAO Doc 9137-An/898, 1st edition, Montreal.

ICAO (1989). *Manual on the ICAO Bird Strike Information System (IBIS)*, ICAO Doc 9332-An/909, 3-rd edition, Montreal.

ICAO (1985). *Airport Planning Manual, Part 2, Land Use and Environmental Control*, 2nd ed.,   International Civil Aviation Organisation, Montreal.

ICAO (1987). *Airport Planning Manual, Part 1, Master Planning*, 2nd ed.   International Civil Aviation Organisation, Montreal.

ICAO (1999). *Environmental Protection. Annex 16 to the Convention on International Civil Aviation, Vol. 1, Aircraft Noise*, 3rd ed., Amendment 6, International Civil Aviation Organisation, Montreal.

ICAO (1999). *Aircraft engine emissions. Annex 16 to the Convention on International Civil Aviation, Vol. 2*, 2nd ed., Amendment 4, International Civil Aviation Organisation, Montreal, July.

Ilkovič, D. (1959). *Fyzika (Physics)*, SVTL Bratislava.

International Civil Airports Association. (1990). *European Airports: their future*, Paris.

Kazda, A. (1992). Airports in the Czechoslovakia faced with the next millenium requirements, STU The 5th ISC, Bratislava.

Kazda, A. (1988). *Vyhodnocovanie prevádzkovej využiteľnosti letísk, práca, (Methods of usability factor assessment)* VŠDS Žilina, Žilina.

Keane, A.G. (1990). *Aviation Security – the Background*, Airport Forum 3.

KIRKLAND, I D and R E CAVES. (1999). *Runway overrun risk assessment*, Transportation Research Record 1662, pages 67-73.

Kotler, P. (1988). *Marketing Management: Analysis, Planning, Implementations and Control*, 7 th ed., Prentice Hall, Englewood Cliffs, New Jersey.

Lemer, A C. (1992). *Measuring performance of airport passenger terminals*, Transpn. Res. A, Vol. 26A, No. 1, pp.37-45.

Lodge, J.E. (1991). *Airport Emergency Procedures*, Loughborough University of Technology, Loughborough.

Loveridge, Denis; Luke Georghiou and Maria Nedeva. (1995). *United Kingdom Technology Foresight Programme*, Delphi Survey, Office of Science and Technology, PREST, University of Manchester.

Maiden, S. (1990). *Know Your Customer Through Passenger Surveys,* Airports Technology International, London.

Manchester Airport, (1999). *Environmental report,* 1998-99.

Mortimer, L.F. (1992). *Ambitious Programme of Future Work to be Undertaken by CAEP,* ICAO. August.

Moutinho, Luiz and Stephen F. Witt. (1995). *Forecasting the tourism environment using a consensus approach,* J. Travel Research, Spring, pp.46-50.

Odoni, A R And De Neufville, R. (1992). *Passenger terminal design,* Transpn. Res. A, Vol. 26A, No. 1, pp. 27-35.

Oliver, J. (1999). *Guiding light,* Passenger Terminal World, Oct., pp 65-69.

Ollerhead, J. B. (1995). *Assessing the impact of aircraft noise upon the community.* Paper presented to conference: Environmental Aspects of Air Transport, Royal Aeronautical Society, London, September.

Orrell, K. (1989). *Airport Commercial Review – a Blueprint for the Future,* Airport Technology International, London.

Pavlíček, M. (1980). *Hluková zátěž na odbavovací ploše, (Noise load on apron).* Letecký obzor 3.

Pearsons, K. S., D. S. Barber, B. G. Tabachnick and S. Fidell. (1995). *Predicting noise-induced sleep disturbance.* Journal of the Acoustical Society of America, 97(1), January, 331-338.

Peen, J., T. Rallis. (1990). *The Traffic Connections Between Airports and City Centers,* Journal of Advanced Transportattion Vol. 24, N^O.3.

Pedoe, N. T., D. W. Raper And J. M. W. Holden. (1996). eds.*: Environmental management at airports - liabilities and social responsibilities.* Thomas Telford, London.

Phipps, D. (1991). *The Management of Aviation Security,* Pitman.

Pilling, M. (1994). *Why Airlines Should Turn off APUs,* Airports International, July/August.

Seneviratne, P N And Martel, N. (1995). *Space standards for sizing air terminal check-in areas,* Jnl. of Transportation Engineering, Vol. 121, No. 2, pp. 141-149.

Shaw, S. (1989). *Airline marketing and Management,* Pitman, 3rd ed.

Schönrock, B. (1989). *The challenges of the 1990´s are airports equipped for them?* ICAA/EUR9/Additional Document, Jerusalem.

Smith, M. J. T. (1989a) *Aircraft noise.* Cambridge University Press, Cambridge, England.

Srnský, S. (1986). *Starosti s leteckým petrolejom, (Problems with aviation kerosene)* L+K 11/86, Praha.

Stadler, G. (1992). *Air Transport and Environment*, Academy Graz.

Steffen, M. (1993). *The Munich Experience,* 2nd ACI Europe Winter Service Workshop, Budapest.

Stefl, B. A. (1990). *Aircraft Anti-icing Fluid Technology and Application,* Airports Technology Internationals, London.

Subramanian, N. V. (1997). *Airport safety problems and solutions,* in proceedings of ACI Airport Operational Safety Conference, Civil Aviation Authority of Singapore, April.

Taneja, N. K.(1979). *The Commercial Airline Industry*, D.C. Heath and Company, Toronto.

Tošic, V. (1993). *Modelling passenger Terminal Operations at Medium-Sized Airports*, The Air Transport in Central Europe Conference – Jasná, VŠDS Žilina.

Trunov, O. K. (1989). *Beating the Aircraft Ground Icing Problem,* Airports Technology International, London.

Tyler, C. (1999). *Airport access for all*, Airport World, Vol. 4, Issue 6, Dec. 1999/Jan 2000, pp 56-59.

Walker, Ch. (1991). *Less Lighting on the Road to Cat 3*, Airports International, October.

Relevant FAA Advisory Circulars:

50/5060-5 *Airport Capacity And Delay* (9-23-83).

150/5070-6A *Airport Master Plans* (6-85).

150/5190-4A *A Model Zoning Ordinance to Limit Height of Objects Around Airports* (12-14-87

150/5200-18B *Airport Safety Self-Inspection* (5-2-88)

150/5200-30A *Airport Winter Safety and Operations* (10-1-91) 150/5200-30A, Chg. 2 (3-27-95)

AC150/5200-30A, Chg. 3 (11-30-98AC 150/5200-30A, Change 4 (11-15-99).

150/5200-31A *Airport Emergency Plan* (9-30-99

150/5200-33 *Hazardous Wildlife Attractants on or Near Airports* (5-1-97

150/5210-2A *Airport Emergency Medical Facilities and Services* (11-27-84).

150/5210-5B *Painting, Marking, and lighting of Vehicles Used on an Airport* (7-11-86).

150/5210-6C *Aircraft Fire and Rescue Facilities and Extinguishing Agents* (1-28-85)

National Fire Protection Association (NFPA) Aircraft Familiarization Charts Manual (3-6-98)

150/5220-10B *Guide Specification for Water/Foam Aircraft Rescue and Firefighting Vehicles* (10-20-97)

150/5220-22 *Engineered Materials Arresting Systems (EMAS) for Aircraft Overruns.* (8-21-98).

150/5230-4 *Aircraft Fuel Storage, Handling, and Dispensing On Airports* (8-27-82).

150/5300-13 *Airport Design* (9-29-89). (Consolidated reprint includes changes 1 thru 5)

150/5300-14, Chg. 1 (8-13-99). *Design of Aircraft Deicing Facilities*, Change 1

150/5320-5B *Airport Drainage* (7-1-70

150/5320-6D Airport Pavement Design and Evaluation (1-30-96).

150/5320-12C *Measurement, Construction, and Maintenance of Skid Resistant Airport Pavement Surfaces* (3-18-97).

150/5320-14 *Airport Landscaping for Noise Control Purposes* (1-31-78).

150/5320-15 *Management of Airport Industrial Waste* (2-11-91). 150/5320-15, Chg. 1 (4-22-97)

150/5325-4A *Runway Length Requirements for Airport Design* (1-29-90). (Consolidated reprint includes Change 1 dated 3-11-91).

150/5340-1H *Standards for Airport Markings* (8-31-99).

150/5340-18C *Standards for Airport Sign Systems* (7-31-91). 150/5340-18C, Chg. 1 (11-13-91).

150/5345-28D *Precision Approach Path Indicator (PAPI) Systems* (5-23-85). Reprint incorporates Change 1

150/5360-9 *Planning and Design of Airport Terminal Facilities at Non-Hub Locations* (4-4-80).

150/5360-10 *Announcement of Availability--Airport Landslide Simulation Model* (ALSIM) (4-24-84

150/5360-13 *Planning and Design Guidelines for Airport Terminal Facilities* (4-22-88).

150/5360-14 *-Access to Airports By Individuals With Disabilities* (6-30-99)

150/5390-2A *Heliport Design* (1-20-94).

INDEX